Failure Point

FAILURE POINT

How to Determine Burning Building Stability

Howard J. Hill

Fire Engineering

Disclaimer

The recommendations, advice, descriptions, and the methods in this book are presented solely for educational purposes. The author and publisher assume no liability whatsoever for any loss or damage that results from the use of any of the material in this book. Use of the material in this book is solely at the risk of the user.

1421 South Sheridan Road
Tulsa, Oklahoma 74112-6600 USA

800.752.9764
+1.918.831.9421
sales@pennwell.com
www.FireEngineeringBooks.com
www.pennwellbooks.com
www.pennwell.com

Marketing Coordinator: Jane Green
National Account Executive: Cindy Huse

Director: Mary McGee
Managing Editor: Marla Patterson
Production Manager: Sheila Brock
Production Editor: Tony Quinn
Book Layout: Publishers' Design and Production Services, Inc.

Library of Congress Cataloging-in-Publication Data
Hill, Howard, 1948–
Failure point : how to determine burning building stability / Howard Hill.
p. cm.
Includes bibliographical references and index.
ISBN 978-1-59370-283-0
1. Fire extinction--Accidents--Prevention. 2. Building failures--Risk assessment.
3. Structural failures--Risk assessment. 4. Live loads. 5. Emergency management. I. Title.
TH9182.H55 2012
690'.21--dc23 2012010154

Printed in the United States of America

2 3 4 5 16 15 14 13

This book is dedicated to the 343 FDNY firefighters who inspired us with courage, hope, and dignity because of what they did on 9/11/2001—men like my cousin, Edward Rall of Rescue 2. He was the type of person you would want coaching your kid's little league team and coming to help you if your life depended on it.

Contents

FESHE Outcomes Correlations

Course requirements for the Building Construction for Fire Protection course developed as part of the Fire and Emergency Services Higher Education (FESHE) model curriculum.

		Chapter			
		1	2	3	4
FESHE Outcome		Page no(s).			
Describe building construction as it relates to:	firefighter safety		51, 64, 65	69–97	135–158
	building codes				125–135, 167–179, 181
	fire prevention	4			116–118, 124–125, 159–160, 174–176
	code inspection				174–175, 179–180
	firefighting strategy	13	24, 61–64	99–100	
	tactics	13–14	18, 21, 24, 30, 63–64	73, 77	165
Classify major types of building construction in accordance with a local model building code.		6–12			
Analyze the hazards and tactical considerations associated with the various types of building construction.		6–13	17–39, 54, 60		137
Explain the different loads and stresses that are placed on a building and their interrelationships.			17–18, 20, 25, 39–50, 52, 56–57, 61		
Identify the function of each principle structural component in typical building design.			17–37, 51		
Differentiate between:	fire resistance				103–111, 114
	flame spread		43		122
and describe the testing procedures used to establish ratings for each.		2	41–42		111–115, 119–121, 123, 181–182
Classify occupancy designations of the building code.		6	17		174
Identify the indicators of potential structural failure as they relate to firefighter safety.		1–16	17–65		103–173
Identify the role of GIS as it relates to building construction.					NA

Acknowledgments

I thank my grandfather, Joe Daly, who started the progression of making me who I am today by becoming a NYC firefighter in 1929, and my uncles, Chief Joe "Mother" Daly and Chief Jack "Black Jack" Daly, who guided me by being who they were. There was never any chest thumping at family gatherings, just men with really good stories about their work.

I am deeply indebted to Chief of Department Peter Gancie, who gave me the opportunity to teach a safety course on building collapse, and to all the chief officers of the FDNY who participated in sharing their experience and knowledge.

The staff at PennWell Publishing made this book better than what it was; I thank them for their efforts on my behalf.

I want to thank my mother for giving me the greatest gift of all by helping me not to hate.

And I want to thank my sons, Luke and Adam, for being Luke and Adam and bringing me the joy of who they are.

I am grateful for an opportunity to thank my wife, Maggie, for being with me every step of the way during this process, providing intelligent, heart-true support for what is written on these pages.

Introduction

I have spent more than 34 years working as a New York City firefighter, just as my grandfather and my uncles did before me. During that time the most challenging situations for me involved knowing when a building on fire might collapse. Initially, like most others I suspect, I thought that every building I was in where flames crept up the walls and went rolling over the ceiling would come crashing down. Day by day, I learned that most buildings could withstand an amazing amount of destruction from fire and still stand. Roofs burned off, stairways crumbled, and water filled the basement, yet the basic building structure remained (fig. I–1).

Fig. I–1. Most buildings can withstand an amazing amount of destruction from fire and still stand (courtesy FDNY).

Tenement buildings appeared to be almost indestructible. Six or seven fires, one after another in the same building, sometimes destroyed all the floors with nothing but rubble inside, yet the exterior walls stood straight and true. How this could occur was difficult to understand.

Then, as I spent more time on a hose line I started to think, why did that ceiling collapse so suddenly? Why did the entire building come down with a roar without any apparent cause? And most importantly, how did that chief know when to pull us out of the building before it collapsed?

It seemed to me to be an unfathomable skill that earned total respect and awe. I couldn't imagine where to begin to look or even what to look for!

When I had 25 years on the job I was promoted to the rank of battalion chief. I experienced many situations directing firefighting personnel into buildings on fire. After a quarter century fighting fires in New York City (NYC), you pick up a few signs that will warn you when something is about to go terribly wrong: strange sounds, overloaded areas in a building, heavy fire damage undermining the strength of a floor, and that soft spongy bounce as you walk on a fire-weakened roof.

I witnessed many buildings fail from fire-induced damage, and countless partial collapses. There were some unusual collapses, one caused by bales of rags expanding from water absorption and pushing out a bearing wall. While learning the job, the tactics seemed the same: always charging in, an all-out effort whether the building was vacant or occupied, rarely hesitating. And who knew, maybe there was a vagrant slouched in a corner to be rescued. This was true dedication—save a life no matter who it was.

I still thrill to that thinking and admire those who showed me what it took to push aside the fear we all had and get the job done. But as I started moving up in rank and went to endless firefighter funerals, I began to think there should be solid considerations and measures to judge before I ordered so many fine firefighters to put their lives at risk. This book tries to describe all that is known to determine when a building on fire will collapse. It includes risk management guidelines to help determine when to risk firefighters' lives and when to let the building burn. And I present some solutions that I hope will help prevent fire-induced building collapses in the future.

I do not offer platitudes such as "firefighters should not be working on any roof structure that is actually burning." Firefighters *will* be operating on these roofs; it is a matter of what type of roof and how much fire (fig. I–2). Nothing is routine; fire officers must make decisions, often based on incomplete information, about what is occurring in a complex situation. This is what we are required to do. This book presents the real-world risk that firefighting entails and the judgments made to protect the public and firefighters in an appropriate manner.

My study of building collapse started back in 1998 when the chief of department, Peter Ganci, asked me to create a safety officer course for the New York City Fire Department (FDNY). I was honored to undertake this assignment, but I didn't know where to begin, particularly when it came to finding information concerning the stability of a burning building. So I started where I learned the most on the job, talking to those who knew more than I did, the acknowledged leaders of the department: Ray Downy, Larry Stack, Pete Ganci, and so many others who took the time to share their knowledge on all aspects of this subject.

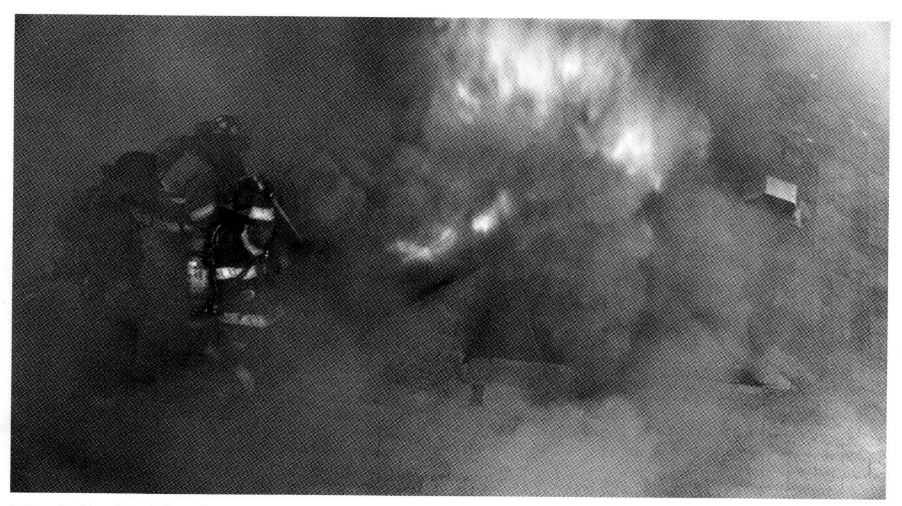

Fig. I–2. Nothing is routine; fire officers must make decisions, often based on incomplete information, about what is occurring in a complex situation.

I thought back and remembered what firefighters and officers showed me, such as when Lieutenant Maurice Walsh grabbed my shoulder as I was moving a hose line nozzle into an apartment fire doorway and said, "When you see smoke like that, wait until that fire blows out the door before moving in." Sure enough, a moment later heavy fire blew out the doorway and then let up. Stay low and let it blow! You don't need to be told things like that twice. You learn quickly on "the job."

The result of that research is gathered in this book, along with information given by the chief officers who took the command safety course that I had the honor to teach. It was always a joy for me to be the one receiving this knowledge from so many people I admired. I am passing this information on to you with the hope that you can be better prepared to make that most difficult of decisions a fire officer has to make: "When do I withdraw members from this burning building?" See figure I–3.

Fig. I–3. When do you withdraw members from a burning building? (Courtesy FDNY)

Fire-Induced Damage to Buildings

1

Overview of Fire-Induced Building Collapse

While researching material on fire-induced building collapse, I reached out to people in the private sector to see what information they had to offer on this subject. I contacted a well-known firm that demolished highrise buildings and asked what they knew. I will never forget the conversation: "Well the first thing you do, Chief, is to take a look at the building's blueprints and find the main load-bearing structural members."

I had to stop him there by saying, "Wait. A chief usually never sees the building before being called to the 3 a.m. fire and has no blueprints or any type of information about the building."

He was amazed. He said, "You mean to tell me that before you fight the building fire you don't know anything about it?" I told him it was a rare thing to have any detailed information concerning the building's structure. And even if you did, you couldn't tell how much damage had been done by the fire before you arrived because you couldn't be sure how long it had been burning, and smoke obscured making an inspection of the interior.

He didn't know what to say and had little to tell me after that. He knew exactly how a building would collapse. He carefully placed explosives to bring down a building only after a meticulous study of all its structural elements.

I discovered that there was a lack of detailed information concerning fire-induced building collapse. A National Institute of Standards and Technology (NIST) study performed in October 2008 on fire testing of materials encountered similar problems when trying to find information on building collapse.

> Difficulties were encountered during this survey in readily identifying news, and other credible sources of historical and technical information on the fire-induced collapses of buildings. The potential data sources were fragmented, often incomplete, and sometimes conflicting. This lack of data and information significantly hampered the development of a more complete understanding of the magnitude and nature of fire-induced collapse. A centralized reliable body of catalogued information on fire-induced building collapses is needed.[1]

I have witnessed many buildings partially and/or fully collapse where no record of the collapse was reported unless there were some extenuating circumstances involved, such as a firefighter's death or serious injury.

Fire Chronicle

In 2005 I was the incident commander at a fire involving a three-story 25 × 60-ft wood frame residential building. The building was attached on one side to a similar building with an empty lot on the other side (fig. 1–1).

Due to heavy fire damage on the lower floors of the fire building, I had removed operating forces and set up a defensive perimeter, protecting the exposed attached buildings and the buildings in the rear with hand lines. I anticipated the potential 90-degree wall collapse and established an appropriate collapse safety zone. After operating for almost an hour, with no warning, the entire building suddenly and completely collapsed in a pancake fashion, leaving a 7-ft-high pile of rubble where a moment ago a three-story building stood. No one was injured, and the fire did not spread from the original fire building.

Fig. 1–1. Three-story wood frame residential building (courtesy FDNY)

My experience told me that with heavy fire on the lower floors in a wood frame building, there was significant potential for this type of building to collapse. With no apparent life hazard and property of little value, there was no reason to risk injury to firefighters. I had studied all there was on fire-induced building collapse and used risk management principles to determine the tactics to fight the fire. The combination of these studies prepared me for the decisions necessary and resulted in a successful fireground operation.

* * * *

However, I never reported the collapse of that wood frame building in my fire report. Why? The FDNY uses the U.S. Fire Administration's National Fire Incident Reporting System (NFIRS). This is the standard fire report used by most fire departments across the country. The database used by this report generally does not include information on buildings that have completely collapsed or suffered partial structural failure. The system does include fields for entering collapse-related fire suppression factors information; however, these fields are not required to be routinely filled out and are used only at the discretion of the reporting department. This means they are rarely filled out. Also, these fields do not capture information that could be used in a meaningful study of fire-induced building collapse incidents.

This seriously limits our ability to understand the nature and extent of fire-induced structural collapses. In fact, there is no single central repository of data on fire-induced structural collapse, either in the United States or abroad.

Collapse events are rarely recorded. The amount of time, expense, and effort required to accurately study why a fire building collapsed prevents most fire departments from investigating the cause, unless a firefighter fatality, multiple civilian casualties, or other unusual circumstances occurred.

A significant problem in collecting this data is that in order to detail how a building on fire was weakened to the point of structural failure, a team of trained personnel is required to observe the burning building. A chief arriving in the middle of the night does not have the resources to provide this level of attention. Without the scrutiny of trained building collapse observers at a fire scene, data collection that can provide a collapse cause and effect is limited. Operating personnel focused on achieving tactical objectives can only go so far in providing the observations needed for a scientific study, which requires instruments capturing data on specific aspects of the event. Fire departments can, however, capture what type of building construction was involved in the collapse and other useful information.

The results of a worldwide survey performed by NIST indicated that a total of 22 fire-induced collapses of buildings four stories or more occurred from 1970 to 2008.[2] I am certain that during my career in the FDNY, spanning from 1972 to 2007, in NYC alone there were significantly more than 22 four-story buildings or higher that collapsed from fire-induced damage (fig. 1–2).

There is plenty of work for the fire service community to do in preventing fire-induced building collapse. Structural collapse statistics are needed to alert fire departments on how different types of building construction may fail and the frequency of collapses.

Fig. 1–2. There has been a failure to accurately record structural building collapse caused by fire (courtesy FDNY).

This data should also be made available to building and fire code officials so they can understand how to create laws requiring fire protection standards that prevent total building collapse from fire damage or other causes.

During my research on fire-induced collapse I hunted for any type of information available. The sources I relied on were the books by Chief Vincent Dunn and Francis L. Brannigan. I strongly recommend that, to keep up with the available information on this vital study, firefighters read these authors' works, articles in fire journals, information from building construction professionals, and related reports from government agencies such as the National Institute of Standards and Technology.

Types of Building Construction

Fires in commercial buildings are more dangerous to fire service personnel than those in residential buildings. This is because *per fire incident* more firefighters are killed in commercial building fires. There are more residential building fires, which cause greater numbers of deaths and injuries to firefighters and civilians, but "per fire incident" more firefighters are killed in commercial buildings.[3]

Classifications of buildings

The *NFPA 220: Standard on Types of Building Construction* list types of buildings based on the material used in the structural members.[4] Fire will cause different types of structural weakness in all of these materials, resulting in various kinds of specific collapse potentials.

Fire-resistive construction. *Fire-resistive* construction is designed to survive a complete burnout according to preestablished hourly ratings of major construction elements and assemblies. This type of construction frequently has spalling concrete collapses.[5] Common building elements consist of galvanized steel studs, concrete floors, steel columns, and beams protected by spray-on insulation and the floor-above Q-decking. Often a vapor barrier on the wall insulation is combustible but will be protected by wallboard. This construction type is used in both commercial and residential buildings of unlimited height (fig. 1–3).

Noncombustible and fire-resistive construction. *Noncombustible construction* uses material that will not support combustion, such as unprotected steel and concrete. The important difference between noncombustible and fire-resistive construction is that while basically the same type of materials are used, the noncombustible construction materials are not fire protected. Steel beams and columns are used along with open web steel bar joists and Q-decking or concrete slab roof covered by a membrane. Without ceiling protection, this type of roof has no fire rating, and when exposed to fire a roof collapse often occurs (fig. 1–4).[6]

Fig. 1–3. Fire-resistive construction (courtesy FDNY)

Fig. 1–4. Noncombustible construction

Ordinary brick and wooden joist construction. *Ordinary brick and wooden joist* construction is basically composed of two brick bearing walls with wooden joists running the short dimension of the building. The joist run is usually 25 ft or less, which limits the width of the building unless there is a transfer or intermediate beam. The roof assembly is constructed in a similar fashion to the floors. The cockloft is considered to be the major inherent weakness to fire. The structural load-bearing members do not have any fire protection in the cockloft and burn rapidly in the open air space. The parapet wall frequently collapses due to these construction features (fig. 1–5).[7]

Fig. 1–5. Ordinary brick and wooden joist construction

Heavy timber construction. *Heavy timber construction* refers to the old factory and loft–type commercial buildings. Heavy timber construction sets limitation on the minimum size, thickness, and/or composition of all load-carrying wood members.[8]

Heavy timber construction eliminates most concealed spaces under floors and roofs and requires the use of approved construction details (fig. 1–6).

Fig. 1–6. Heavy timber construction

The elimination of concealed spaces coupled with the inherent fire resistance of heavy timber makes these structures less prone to collapse than other wood frame constructed buildings. However, once fire involves a large area of the building, it will produce intense heat conditions that are very difficult to suppress and control. When this occurs the floor generally collapses first, initiating a wall collapse.[9] The heavy timber buildings being built today often use glued laminated timber as the major load-bearing structural members.

Wood frame construction. *Wood frame construction* uses relatively lightweight wooden structural members. Commonly, in wood frame construction, the primary structural elements in the bearing walls are 2 × 4s. A wall collapse is the most common structural failure.[10] There are several types of wood frame construction:

Braced frame construction, also known as *post and beam construction*, was commonly used in the early 1700s. These buildings are now nearly 300 years old, and their wooden structural members are often rotted. The mortise and tenon joint connections are the weak spots. Often collapse occurs suddenly in an inward/outward manner, where part of the wall will fall into the remaining structure and part will fall outward. The weakened wall cracks horizontally, and the weight above pushes down, kicking the lower portion of the wall out. The upper part buckles in, crashing down on the interior, resulting in the collapse of the floors. This collapse frequently occurs with no warning.[11]

Balloon framed construction started in the 1800s and continued until the 1940s. This type of construction reduced the amount of weight in a wall, hence the phrase "so light the wall went up like a balloon."[12] The studs run the entire height of the building from the foundation to the roof eaves. A ribbon or ledger board connected to these walls carries the floor joists on each floor. The 90-degree wall collapse is most likely due to the long studs on the exterior walls running from the foundation to the roof. The wall generally collapses outward in one full section, creating a large collapse zone. A sign of an imminent collapse of this wall is when the corners start to split away from the adjoining walls.[13]

Platform wood frame construction became common after 1940 and is the least likely to collapse of all the wood frame construction types. Its unique characterization is having each level built on top of another level, usually limited to three levels, with voids limited to one floor. This type of construction is still being used today (fig. 1–7).

Fig. 1–7. Platform wood frame construction (courtesy FDNY)

Wooden truss construction uses a parallel wood truss system primarily as beams, but they can also be used as columns. The truss system is engineered to use lighter structural elements with less mass, which significantly reduces its load-bearing capacity when subjected to the effects of fire. The top chord is under compression, the bottom chord is under tension, and the web connects the two chords and distributes loads.

A truss joist floor system creates an enclosed void space once wall board is installed. It is open end-to-end and is used to run electrical; plumbing; and heating, ventilation, and air conditioning (HVAC) systems. This void provides an open space for fire to travel and ignite the exposed wooden joist structure. Once this area is penetrated, fire will spread rapidly, and early collapse can be expected.

Fires burning in the parallel chord truss void often show a band of smoke on the exterior at the floor level. The combination of rapid fire spread and lack of fire resistance of the wooden truss potentially results in early fire collapse with no warning signs.[14]

Wooden truss open areas can also allow the spread of fire up to the floor above when construction methods rest the bottom chord on the bearing wall (fig. 1–8).

Fig. 1–8. Fire spread area at wooden truss bearing wall connection point (courtesy FDNY)

Wooden I-beam construction uses a wooden web and 2 × 4s as the top and bottom flanges. They are prone to early failure if the beam is exposed to active burning. When glue is used to connect the web to the flanges, it can rapidly lose its ability to adhere to the structural material when heated in the range of normal fire temperatures. These beams can be as long as 60 ft, which will affect a large number of other structural elements if they fail. They are generally spaced at 32 in. on center. The expected failure time is less than 5 minutes when actively burning.[15]

The char rate that wood typically develops is 1/40th of an inch per minute. After approximately 30 minutes of being subjected to flame, ¾ of an inch of each surface will be burned.[16]

The burning characteristics of wood are:

- Wood ignites at around 480°F and begins to char at around 300°F.
- Wood retains some of its original strength when subjected to fire. The initial strength lost is due to surface charring.[17]

Today, many types of new wood construction materials are combined with lightweight metal materials, resulting in buildings with a high potential for early collapse when the load-bearing structural members are weakened by fire.

Construction types can be further grouped into two general classifications: *unframed construction*, where the structure depends on bearing walls to support the roof and floors, and *framed construction*, where the primary structural members consist of columns and girders that support the building. Generally, collapses involving framed buildings are more localized compared to unframed buildings. This is because in unframed buildings, if one wall collapses it can bring down several floors, whereas in framed construction, if a section of the frame fails it will generally affect only the floor it supports.[18] However, as was witnessed in the World Trade Center Building 7 (7 WTC) collapse, framed construction is not immune to total collapse from fire-induced damage (fig. 1–9).

Fig. 1–9. The WTC tower buildings were framed construction (courtesy FDNY).

Structure Dictates Firefighting Tactics

The importance of understanding the construction characteristics of different building types is not only necessary for a proper evaluation of the building's stability; it is also required to determine what actions are being performed by firefighting personnel.

What is the prime factor in deciding how firefighting forces are deployed? Fireground tactics are determined by *structure*. Is the building large or small, highrise commercial or one-family residential? What is it made of and how is it made? These, in combination with the existing life hazards, are the factors that will determine how a fireground commander uses the firefighting personnel at the scene.

The FDNY has specific operational procedures for several construction types: *private dwellings* (one- and two-family homes), *brownstone buildings* (three- or four-story residential buildings), *Queen Anne buildings* (large, older wood frame construction), *multiple dwellings* (tenements and highrise apartment buildings), *taxpayers* (strip mall commercial buildings), and *highrise commercial office buildings*.[19]

For each of these different construction types there are tactical procedures based on the different building features: where to stretch the first hose line, what areas are searched initially, and by whom, and so on. Within each set of tactics there are variations based on structural characteristics, such as roof design types. For example, a flat roof will require a different set of actions than a peaked roof. The type of material used in construction will also decide how members operate; a building made from fire-resistive materials will require different procedures than buildings constructed of wood.

The architectural design of a building and its fire protection features influences firefighting tactics and standard operating procedures (SOPs). The building's occupancy and construction characteristics determine where and how hose lines are stretched, where searches for life are performed, and how to keep the fire from extending to other areas of the building.

The impact of firefighting operations on a structure is another consideration that should be taken into account to help determine any weakened structural stability. The force of hose streams laterally striking walls, removing plaster and sheetrock to check for fire extension, and other such activities create disturbances to the load-bearing weight distribution in structural elements. The knowledge of where firefighters are inside the building is valuable information pertaining to ongoing damage to structural elements at specific locations.

References

1. NIST GCR 02-843-1 paper, Analysis of Needs and Existing Capabilities for Full-Scale Fire Resistance Testing, 2008.
2. Ibid.
3. NIST 7069: Trends in Firefighter Fatalities Due to Structural Collapse, 1979–2002.
4. NFPA 220: Standard on Types of Building Construction, 2012 ed.

5. Dunn, Vincent. *Collapse of Burning Buildings: A Guide to Fireground Safety, 2nd Ed.* (Tulsa: Fire Engineering Books & Videos, 2010), 216–217.
6. Ibid., 50–51.
7. Ibid., 51.
8. Brannigan, Francis. *Building Construction for the Fire Service, 4th ed.* (Sudbury, MA: Jones & Bartlett, 2009),129.
9. Dunn, 51–52.
10. Ibid., 204.
11. Ibid., 204.
12. Brannigan, 96.
13. Dunn, 204.
14. Fire Department of New York. Safety Message 90: "Truss Construction."
15. Brannigan, 551.
16. American Society for Testing and Materials, ASTM E 119.
17. The American Institute of Timber Construction, Technical Note 7, Jan. 1996.
18. Brennan, Tom. Signs of Impending Building Collapse. *Fire Engineering*, July 2000.
19. FDNY Firefighting Procedures. DCN: 1.05.06.

An Analysis of Structural Damage 2

What to Look For and Where to Look for It

Think of the fire building as a living person. There are three critical stages of a person's life: birth, sickness, and death. A building is most unstable during initial construction, major alterations, and demolition. A high percentage of fire-induced building collapses occur in buildings that are in one of these three stages.

The building's skeleton consists of the foundation, load-bearing walls, beams, columns, the floor system, the roof system, and the connections between these elements. To carry its own weight (dead load) and all the other loads and forces being applied, the building elements and connections are subjected to various forces: tension, bending, shear, torsion, and compression.

Buildings are primarily designed to resist the vertical forces from gravity. The roof and floor systems carry these vertical forces to the supporting beams. The beams carry the forces to the columns and bearing walls, which then carry the forces down to the foundation and the supporting soil. The progression of carrying these forces from the roof down to the soil is the load path. The failure of any load-bearing structural member or connection along the load path can lead to building collapse.[1]

The order in which structurally damaged elements should be evaluated is

1. Bearing walls
2. Columns
3. Girders and beams

Inspect the major load-bearing elements to determine prime weakness, and then check for partial collapse of critical supports and unsymmetrical loading. If there are no obvious structural concerns, generally the evaluation of the building can start at the foundation, because this area affects the entire building.

The following analysis provides information to recognize the identifiable collapse dangers of different construction materials and where to look for specific warning signs that indicate structural weaknesses.

Walls

Walls transmit loads to the ground in compressive force. The intersection point of two walls tends to be the most stable point, because this is where they act as a brace to each other against asymmetrical loading. Walls are designed to accept vertical compressive loads. Any type of lateral loading often initiates a wall collapse.

Load-bearing walls carry a load of some part of the structure in addition to the weight of the wall itself. When the weight of the roof is used to add compression the force acts as a monolithic brace. (The monolithic style of construction is very stable and is characterized by a structure that is carved, cast, or excavated from a single piece of material.)

The collapse of a bearing wall often brings down a significant portion of a building and is one of the leading causes of catastrophic building collapse. Any extinguishing or overhauling activity that might damage the integrity of this wall must be closely supervised and monitored.[2] Failure of this wall will often result in a progressive collapse of other structural members.

Masonry walls

Masonry bearing walls should be checked for cracks and bulges. A brick bearing wall can be identified by having every seventh course set on end. These bricks are set on end to act as a brace for the stretchers, which are laid horizontally. When bricks are not arranged in this manner, it indicates a brick veneer wall that is attached to other load-bearing wall members.

Exterior metal braces on a brick wall are used for different purposes: they can be used to provide stability by fastening the wall to the floor, making the wall more rigid. Braces used in a regular pattern indicate a decorative feature. When these braces are used, prepare for the worst, and assume there is a structural weakness in the wall (fig. 2–1).[3]

Cracks in brick walls usually indicate foundation problems, which may be serious. Be aware that stucco is often used to hide the deteriorated condition of bricks. Sand lime mortar, used in older buildings, creates a potential failure when it is washed out due to weather and/or exterior firefighting water stream operations (fig. 2–2).

Fig. 2–1. Exterior metal braces on a brick wall

Fig. 2–2. Deteriorated sand lime mortar brick wall

Any cockloft fire in a large, wood and brick, one- or two-story building with a parapet wall must be suspect for possible collapse of the wall. If the parapet wall is loaded with signs, decorative facades, or overhead security doors, this creates additional loads that were not designed into the strength of the wall, and the odds of early collapse are increased greatly.[4] The likely point of initial collapse is the highest section of the parapet wall because of the increase of unsupported wall area.

When a single column or beam gets knocked away from supporting a heavy masonry wall, it causes a load shock to the other supporting members that can initiate a progressive collapse of several structural steel columns or beams. The more interconnections the wall has, the more stable the wall and the more resistant it is to load shock. Strong structural interconnections increase the overall building strength and stiffness by allowing all of the building elements to act together as a unit. Inadequate connections represent a weak link in the load path of the building, making the wall more susceptible to failure when lateral load forces are applied. Bearing walls gain more stability from the weight applied on them than do nonbearing walls and benefit in strength with interconnections.

The *corner area* where the walls meet is a good area to assess the stability of the structure. If a separation between these walls develops, there is a very strong probability that a collapse is imminent. However, even this significant warning sign does not indicate that a *definite* collapse will occur.

What actions should be taken when collapse warning signs are present?

Fire Chronicle

I was the incident commander at a multiple alarm fire involving a one-story ordinary brick construction commercial building, 150 × 200 ft. Fire had full possession of the interior with flames shooting up through the skylights that had cracked and broken from the heat. I ordered a defensive strategy with lines on the roofs of adjoining buildings and surrounding properties. Members and apparatus were positioned outside the collapse zone and large-caliber streams were put in operation.

Approximately 20 min after arrival, in one corner area of the fire building, the walls began to split 10 in. apart at the roof level, and the split tapered down 4 ft to where the walls again joined together. I informed all officers in the area that this was a sign of imminent collapse and to closely monitor the crack between the walls (fig. 2–3).

Fig. 2–3. Separation at a building's corner areas often indicates significant structural weakness and can precede complete wall collapse.

After the outside streams had extinguished most of the fire, I performed a perimeter inspection. The walls were still standing, and the crack between them remained the same as it was when I first discovered it.

* * * *

Why didn't these walls collapse? Most of the roof had burnt through and there were several areas where local collapses of interior walls had occurred, shifting laterals loads to sections of the exterior walls, but all four exterior walls stood.

I am describing this incident to show how unpredictable structural collapse can be. Sometimes the building crumbles before your eyes with no warning whatsoever. Other times, major portions of the building fail, yet the exterior walls and substantial portions of the building stand. The important guideline to always follow is: *if warning signs indicate collapse is imminent, withdraw firefighting personnel from the area*. No one can say with absolute certainty exactly when a fire-damaged building might fail. It is a judgment call.

There was no life hazard in the building, and the property had been destroyed to the point that it had little or no value. Risk management techniques were used to guide my decisions to order an outside, defensive strategy. Though the building did not collapse, these were the appropriate actions to take.

The debris from a masonry building collapse weighs 125 to 150 lb per cubic foot.[5] This should be used to calculate the additional loads being put on a structure when there is a localized collapse inside the building.

Concrete walls

Tilt-up concrete construction. Tilt-up concrete construction is typically one to three stories in height and consists of multiple monolithic concrete wall panel assemblies. The structure can also use an interdependent girder column and beam system for providing lateral wall support of floor and roof assemblies.

Nonbearing tilt slab walls are connected to the side walls and the roof assembly. If fire is weakening the roof connection points, there is a danger of a 90-degree wall collapse.

Polystyrene form blocks. Besides the standard concrete reinforced wall, in which the concrete is poured into forms at the job site, there is another type of wall that uses polystyrene foam blocks as the forms for a reinforced concrete wall. Light reinforcement rods are woven into the forms where the concrete is poured.[6] The polystyrene blocks stay in place after the concrete has been poured and become part of the structure. Galvanized steel attachment strips, located on each side of the forms, allow metal screws to attach any type of wall covering and interior studs. The polystyrene foam will not burn but will create smoke. These walls create a tightly sealed interior, which can add to a backdraft potential when the front door is opened during the initial attack on the fire (fig. 2–4).

Reinforced concrete walls generally maintain their stability better than combustible materials when exposed to fire.

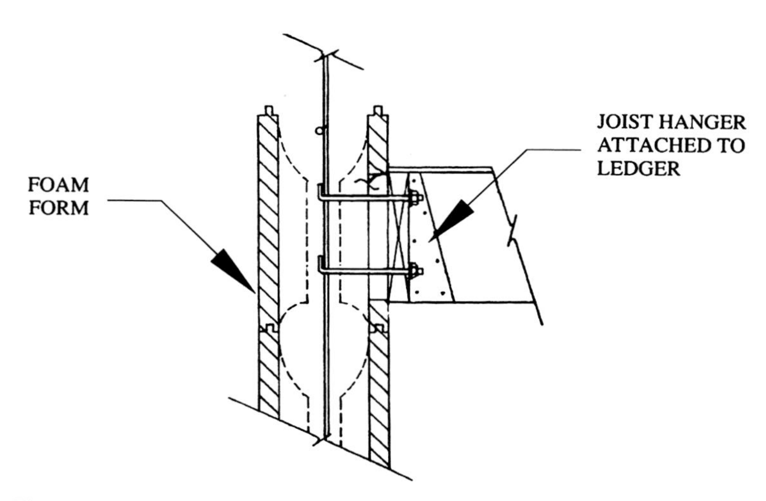

Fig. 2–4. Concrete wall made with polystyrene form blocks

Wood frame walls

The bearing walls of wood frame construction should be identified early on and watched carefully during the fire. The longest wall is usually the bearing wall, and the longest roof ridge rafter often runs parallel with the bearing wall. From the interior of an older wood frame building, a bearing wall can be identified by looking at the run of the wooden subfloor planks, which run parallel to the bearing wall.

When a wood frame bearing wall fails it will often cause floor and/or roof collapse, which frequently results in a total collapse of the structure.

Overloaded wood bearing walls with additional loads that are not self-supporting, such as signs, fire escapes, and brick veneer, are prone to early collapse. There are three major considerations when judging the potential for a wood frame building to collapse from fire destruction:

- Location. Serious fire on the lower floor affects the load-bearing capability of 2 × 4-in. studs and is more likely to cause a building collapse since they support more of the structural load of the building. Upper floors are not subjected to this additional load stress.
- Time. The longer the structure has been exposed to flame, the greater the possibility that the bearing walls have been weakened.
- Size and stage of fire. Collapse is most likely to occur during the fully developed fire stage and/or the decay stage immediately following it, because the building has suffered its greatest damage during these two stages.[7]

The general rule is: the larger the fire, the greater the destruction, and consequently the greater the collapse danger. However, the degree of collapse danger depends on what the fire involves, the contents, or the structure.

Content fires rarely cause collapse in and of themselves; they contribute to the collapse once the structural elements are affected by the contents on fire.

Decisions made regarding the collapse potential of the fire building require a determination of whether or not the *structure* is burning. A fire in a corner location wood frame building, where two-thirds of the structural members are involved in fire, should be considered in danger of imminent collapse. This is because the primary structural members in a wooden building are losing their load-bearing capacity as they burn, and a corner building lacks lateral support.

The *time* the fire is burning, the *size* of the fire, and the *structural damage* that is continually occurring must all be factored together to accurately estimate the collapse danger. Time is not on your side. In order to address the possibility of losing track of time while operating, the FDNY dispatcher periodically alerts the incident commander how long units have been on the scene to ensure that the burn time information is constantly being monitored. An incident commander should continually factor in the amount of burn time with the progress of the interior attack and the type of building construction that is being weakened. If there is no one else assigned to monitoring elapsed time on the fire scene, then the incident commander must do so by keeping a record of segments of time, 10 to 15 min each, during active firefighting operations.

If the fire is not being progressively extinguished, firefighting forces are losing the battle, and the potential for collapse of the structure is increasing. Any stalled extinguishment efforts or delays with the aggressive attack greatly affect decisions to change strategy from an interior to an exterior operation. The following are warning signs of possible collapse of a wood frame building:

- Exterior shingle and or clapboards that may twist or pop loose
- Damage to the foundation creating unsupported building loads
- Windows and/or doorways distorted

As with masonry walls, if a corner of a wall of a wood frame building starts to split away from adjoining walls, this is a sign the building can collapse at any moment. The weight of wood frame debris is 10 to 20 lb per cubic foot.[8] If there is a partial collapse, this additional load factor on the supporting structural members will be an additional concern.

Columns

The essence of a column is in its ability to carry a compressive load in a straight path. Different types of columns fail in a different ways: piers are short, squat columns that fail by crushing. Long slender columns fail by buckling in an S shape.[9] The longer the column, the less load-bearing capacity it has. As the height of a column decreases, its load-bearing capacity increases.

Columns made from fire-resistive materials

The material that makes up the column affects its strength and performance when heated. Once the yield stresses are reached in a column, there is little reserve strength left, and the column is on the verge of total collapse. The failure of a column may precipitate the collapse of the entire portion of the building dependent upon that column.

Cast iron columns

A cast iron column's fire resistance has been characterized as failure prone due to the brittle nature of cast iron and the questionable casting process, which can create variable thicknesses in the wall of the column, thus allowing the column to heat up at different rates in different areas. However, there is also evidence that cast iron columns can withstand considerable heat and that their main weakness is their gravity connection points.[10]

Gravity connections in a building simply depend on the weight of the building to hold the column in place (fig. 2–5).

If there is a shift in the designed axial loading of a column, the gravity connections can fail resulting in a complete collapse of the structure. *What will collapse if a column fails?*

Fig. 2–5. Cast iron column gravity connection point (courtesy FDNY)

Fire Chronicle

On June 17, 1972, nine firefighters from the Boston Fire Department died in the collapse of the Vendome Hotel. An investigation revealed that, around 1890, alterations were made to the first story. A bearing wall was removed, and dual 15-in. wrought-iron beams carried by a 7-in. cast iron column were installed. These beams and column carried the loads of all the floors above. Later on another bearing wall was removed which carried the loads of all floors above it, and this load was also tied into the dual 15-in. beams carried by a 7-in. column that had a gravity fit connection. As a result of these renovations, the 7-in. column became the main support of the concentrated load at one central point. It is believed that, as a result of the damage caused by the fire, a shift in the load caused column failure, resulting in the collapse of the entire building (fig. 2–6).[11]

Fig. 2–6. When one column fails, the loads that were carried shift to adjoining columns, which can cause progressive collapse (courtesy FDNY).

* * * *

Steel columns

Steel is commonly used as a column material. When steel is subjected to heat above 1,300°F, it may fail. Steel elongates when heated. A 100-ft-long piece of steel when heated to 1.000°F can extend 9 in.[12] This elongation can sever bolts and break welds at connection points, causing a progressive collapse of other structural members (fig. 2–7).

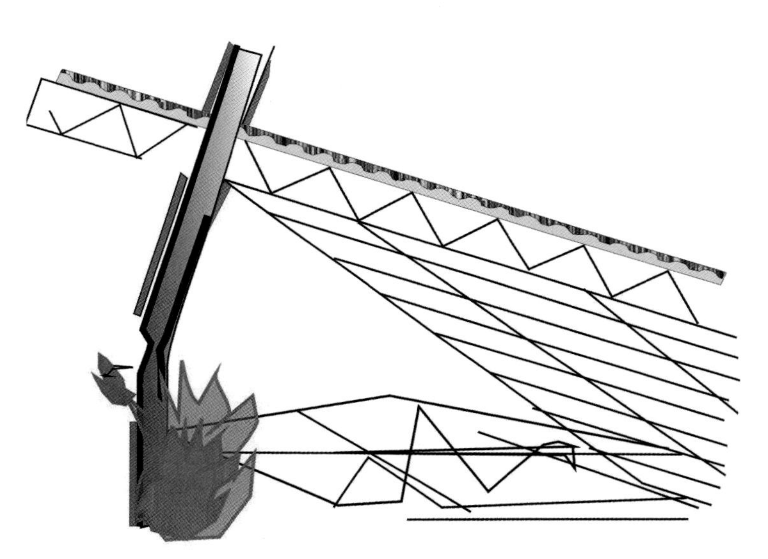

Fig. 2–7. Steel columns may fail when heated to about 1,300°F.

Concrete columns

Reinforced concrete structural materials have specific collapse warning signs. They are

- Damage to the column
- Diagonal cracks
- Structural components shifted
- Exposed rebar or tensioning cable

- Cracked flat slab
- Glass in windows cracked because the columns are out of plumb and the loads are deforming the shape of structural elements
- Spalling

Unsound concrete may be colored and friable. Sound concrete will give a distinctive ring when struck with a hammer.

In concrete columns, if the unity of the composite fails by loss of concrete as a result of spalling from a fire, the steel rods may buckle under the load. Buckling will cause the rods to protrude from the column and indicates significant weakness of the column. Collapsing reinforced concrete will weigh 250 lb per cubic foot.[13] This is a considerable load of general construction debris and warrants close attention in the area of a partial collapse.

Wooden columns

The most significant factor regarding the fire resistance of a wooden column is mass. Studies on square wooden columns have shown that with column size increase, the fire resistance increases sharply. Also, the conduction of heat in wood is directly related to its density. Woods with low density have the highest thermal insulating value, because wood contains a high proportion of cell cavities. In dry wood, these cavities are filled with air, which is one of the best-known thermal insulators. All strength properties decrease as moisture content increases.[14] Once one structural element fails it can often cause load stresses that initiate another structural failure, resulting in the "progressive collapse" effect that can cause a widespread breakdown of the structural supports in a building. This is a typical type of event that occurs when a column fails. Any indication of imminent column failure should be cause for consideration to immediately clear the entire building of personnel.

Floor Collapse

There are specific warning signs of potential floor collapse. Look for a clean, bright, new-looking floor finish against a wall. This indicates that the floor is pulling away from the wall support area and/or is overloaded and is a significant collapse indicator. At least 2 in. of the beam end should rest on the masonry wall or girder (fig. 2–8).[15]

A masonry wall may support a floor joist by having the joist end recessed into the wall or by resting on the corbel shelf. Collapse of a joist end recessed into the wall can create a lateral load stress and additional weakness in the wall. If the joist end rests on a corbel shelf, the collapse damage may be more local in nature and will cause less damage to the bearing wall compared to a recessed supported joist (fig. 2–9).

Any bowing or unsymmetrical appearance is an indication that the floor is being subjected to nondesigned load forces that can result in a total floor collapse.

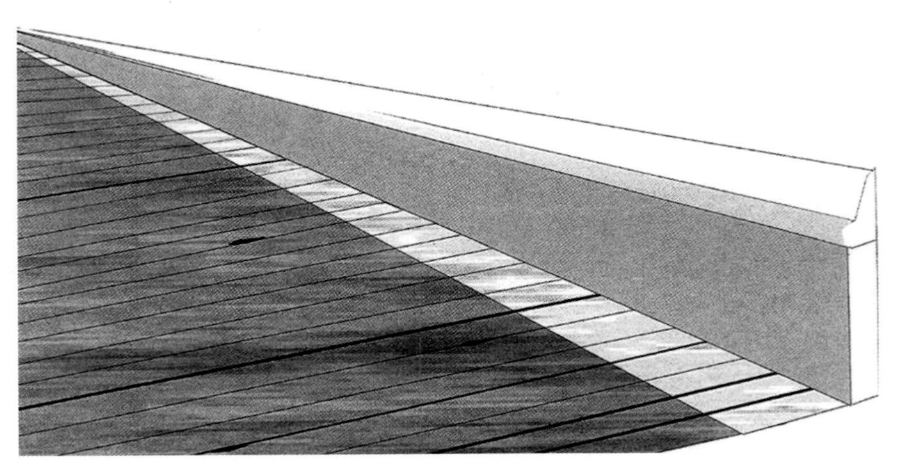

Fig. 2–8. A clean, new-looking floor finish against a wall indicates the floor is pulling away from the wall.

Fig. 2–9. Wooden joists recessed into a brick bearing wall

Wooden I-beams

Wooden I-beams are composed of solid or laminated flanges on either side of a web. The wooden web is often made of composite wood or oriented strand board (OSB), and can have cutouts for utility installations that further weaken its resistance to failure. A beam gets its strength from material mass and its geometry. The lower cost of using geometry (substituting shape for mass used in truss floor joists and I-beams) over mass increases the collapse potential when these structural members are subjected to fire. When there is less mass to absorb the load within the fire-weakened member, it makes a structure prone to collapse.

Figure 2–10 shows a residential alteration construction method using lightweight metal (steel) joists connected to wooden I-beams. The decrease in mass of these load-bearing structural members creates a connection point with little resistance to fire damage.

Fig. 2–10. Lightweight metal studs supporting a wooden I-beam (courtesy FDNY)

Terrazzo floors

Terrazzo floors insulate and hide fire burning below. They are made of concrete and polished marble chip tile. A 4-in.-thick slab of concrete adds nearly 50 lb per square foot to the load the beams are carrying. They are often found in bathrooms, kitchens, and hallways. The construction technique of cutting into the supporting beam, in order to keep the 4 in. of concrete level with the floor, results in further weakening of the supporting beams (fig. 2–11).

Fig. 2–11. Terrazzo floor

Metal C joists

Metal C joists are cold-formed steel structural members that require little or no fire protection. They can support 4-in. concrete floors as well as wood flooring. These C joists can fail without warning when subjected to the heat of a common house fire (fig. 2–12).[16]

Fig. 2–12. Steel C joist (courtesy FDNY)

Precast concrete slabs

Precast concrete slabs are made in the factory where steel strands are stretched and then concrete is poured over them. The concrete sets and bonds with the tensioned steel, putting it under compression and allowing it to span long distances. The slabs are delivered to the job site in precut lengths with openings made to accommodate utilities. Metal plates can be molded into the slabs and then bolted to steel girders. The slabs are joined together at the sides with a tongue and groove joint that is grouted.

The collapse potential at a connection point occurs if a slab is being pulled away from a heated steel girder that has been distorted by fire. Heavy spalling can weaken the slabs to the point of collapse because the reinforcing cables are located in the lower portion of the slab, which is the ceiling area that is subjected to higher temperatures during a fire.

The principal weakness of this type of floor construction is poor connections between floor, wall, and/or roof. The collapse warning signs are

- Broken or twisted connecting hardware
- Cracked connection welds
- Badly cracked walls
- Spalling
- Diagonal cracks in concrete beams

When precast concrete slabs are present, an interior inspection of the beams and/or the corbel shelf is necessary along with the exterior examination to properly assess the stability of this structural element. Look for any significant shifting of structural components.

Pre-stressed hollow core concrete slabs are another type of concrete slab in which the concrete material is removed from the center area, creating less mass, hence less fire resistance and an increase in collapse potential (fig. 2–13).

Fig. 2–13. There is less concrete in hollow-core slabs (courtesy FDNY).

Roof Collapse

Flat roofs

Flat roofs require several considerations when judging stability:

- Type of roof support system; e.g., solid beam, parallel chord wood truss, open web steel bar joist, and so forth
- Presence of a built-up structure on the roof
- Condition of the roof connections to the supporting wall

Open web steel bar joist roofs

With open web steel bar joist roofs, the connection points can break down if any load stresses are affecting the supporting wall. Indications that some breakdown in the capacity to support designed loads can be detected by looking above suspended ceilings for joists out of true shape. Look where the bar joists meet the wall for any signs of twisting or separations. Check the supporting masonry wall for cracks.

Open web steel bar joists are made of noncombustible materials and can span up to 60 ft and be spaced as far as 8 ft apart. Standard fire resistance tests of unprotected steel open bar joists reached a failure point at 1,200°F in 6 to 8 min.[17]

The bowstring truss

The bowstring truss roof is recognized by its hump shape. This roof type is often hidden from street view by parapet walls that extend above the roof area.

Often constructed of heavy timber or metal, the trusses can be spaced 10 to 20 ft apart. Failure of the truss at the ends of the building has caused front or rear walls to blow out forcibly. This is a result of the hip rafters on the end walls hinging out, which requires increasing the collapse zone. When subjected to fire, this type of roof has a history of sudden, catastrophic collapse.[18]

What is the approximate burn time before a large wooden bowstring truss will collapse? An analysis of the Waldbaum fire in Brooklyn, New York on August 3, 1978, revealed that the 3 in. wooden roof truss member was exposed to a fully involved fire in the roof cockloft. The burn time estimate was approximately 35 min, which caused the truss to fail, resulting in the roof collapse (fig. 2–14).[19]

If a serious fire involves the roof portion of a wooden bowstring truss roof, firefighters should not be committed to operations on the roof area.[20]

Note: if immediate evacuation of a bowstring truss roof is required, it is safer to go in a direction toward another truss member than to travel toward a wall that is supported by truss with fire impinging on it (fig. 2–15).

Fig. 2–14. Waldbaum fire, Brooklyn, New York, 1978 (courtesy FDNY)

Fig. 2–15. Bowstring truss roof emergency evacuation toward adjoining truss (courtesy FDNY)

Wooden truss peak roofs

Wooden truss peak roofs are built with reduced mass that will not support any additional loads on the roof, such as firefighters cutting the roof. Generally they have no fire protection and are prone to rapid fire spread and failure (fig. 2–16).

Additional loads on the roof should be part of the initial roof stability size-up:

- Is there any heavy equipment on the roof?
- Are there live snow or water loads?
- Has fire entered the cockloft?
- What type of construction is present?

Fig. 2–16. Wooden truss peak roof

Cell phone roof antennas are common and place additional weight on the roof. The roof structure was not designed to carry this load, and sometimes structural steel supports are installed when heavy machinery is present. Any fire directly under these installations should be constantly monitored for weakness.

Membrane roofs

A fire involving a membrane roof can rapidly spread over the entire membrane material roof surface. They are very fast-moving fires and are difficult to extinguish. A liquid tar fire can drop down and burn the supporting structure, which may weaken load-bearing members and initiate a possible collapse (fig. 2–17).[21]

Fig. 2–17. Membrane roof (courtesy FDNY)

General Causes of Collapse

Some of the general causes of building collapses are

- Lateral loads
- Impact loads
- Heating of steel
- Water absorbent materials
- Water/snow load
- Vehicle striking building
- Backdrafts
- High winds
- Vibrations

- Explosions
- Structural weaknesses

As all homeowners know, water is one of the most destructive natural causes of decay to a structure. Moisture-prone areas, such as connection points where water can accumulate, invite termites and dry rot to further weaken critical structural areas.

Types of Loads and Stresses

An understanding of how structural members are affected by the various forces caused by loads in a building is essential to understanding collapse potential. The interaction between loads, the forces they exert upon construction materials, and the manner in which they are applied determine the stability of a building.

A load is considered an external force that creates stress on a structural member. The action of the load results in the reaction of different types of stress. A load applied to a brick bearing wall by a floor joist puts the wall in compression. The same holds true for columns. When a load hangs from a structural member, it creates the stress of tension. When structural elements are put under stress, they deform. Engineers calculate the amount of deformation by measuring stresses in pounds per square inch, which will cause a percentage of compression or elongation in a particular material.

The effect that these internal and external forces have on structural members is compounded by nominal loads. Nominal loads are defined as the degree of structural movement caused by live loads such as wind, snow, or earthquake.[22] This external force is calculated in the "working stress design" during the planning phase of the building, which determines stresses produced by nominal loads that do not exceed specified allowable limits and contain a factor of safety. A load factor accounts for deviations of the actual load from the nominal load and for the probability that more than one extreme load will occur simultaneously.

Braces (structural members placed diagonally between load-bearing walls or columns) resist both compression and tension forces and are often used to counter wind nominal loads by making the structure stiffer.

Permanent loads are forces with rare or a small magnitude of movement over time, such as the weight from building materials. The product of a nominal load and a permanent load force is a factored load. Factored loads are specified for the strength design, also known as the load and resistance factor design methodology.[23]

These loads are also computed to determine the allowable elastic stress design of structural members and will also contain a factor of safety.

Engineers perform a structural analysis using formulas taking all these forces into consideration to determine if the structure can safely support the weight and the forces applied to it.

Fire damage nominal load

Currently, many fire resistance tests are stopped when certain limiting acceptance temperatures for the materials are reached without any insights into the ultimate structural fire endurance time and failure mode.[24] This gap in understanding the nominal load caused by fire damage is being criticized by the National Institute of Standards and Technology (NIST) and others who voice the need to determine the engineering factors that recognize the effects of structural fire damage. The effect of fire damage is not included on the engineer's or architect's list of nominal loads and should be studied and accounted for in the building's planning stage, the same way an earthquake is.

The effects that nominal loads, wind, and earthquakes, have on the structure is determined, in part, by laboratory tests of scaled models allowing for modeling effects and differences between laboratory and field conditions. The nominal effect that fire has on a structure is not performed at the design level. We would have safer buildings if the effort to calculate fire damage was performed, but for now we must look elsewhere.

Calculations performed on the nominal load from earthquake damage can be considered the closest comparison to fire damage. Information on what is currently being discovered regarding the effects that earthquakes have on buildings will be considered later.

A load force in a building can be either a designed load or a nondesigned load. When determining a designed dead load (permanent load), a conservative estimate of the actual weights of all materials and construction are used to calculate the required values for the structural design. A designed load is one that the builder had anticipated and planned for.

When a load force applied to the load-bearing elements of a structure was not calculated into the design, it is a nondesigned load. Individual structural building elements are not designed with consideration of their performance under fire conditions. A structural member weakened by fire can redistribute loads in unpredictable ways.

Incident commanders must understand the implications of this fact. There is significant difficulty in making accurate judgments of specific failure characteristics of a building without design information concerning the probable manner in which the structural elements will react to the stresses of fire-induced damage.

This does not mean that there is no recourse for analyzing fire-induced building collapse. By examining the characteristics of structural materials that are stressed by fire damage, it is possible to qualify and quantify the factors involved in order to operate safely. The lack of related scientific studies calls for firefighters to supplement their knowledge of how a building might collapse by consulting fire service publications, engineering journals, and other professional construction literature as part of their continuing education of this crucial subject.

Fire load

Fire loads refer to the potential fuels in a building that can support a fire. Buildings made of combustible materials become part of the fire load.

The fire load is measured in British thermal units (BTUs). Different materials generate distinct characteristics when heated; for example, 1 lb of plastics is measured as 2 lb of wood in BTUs.[25] It is undeniable that people in the United States today have more "stuff" in their homes than they used to. This stuff generally consists of plastics, creating a very high fire load compared to home furnishings in the past. Plastics make for hotter fires that spread rapidly and increase the potential for smoke explosions.

Live load

Snow on a roof is considered a live load. The debris from a localized collapse within the building becomes a live load. An additional danger of live loads (also known as nominal loads) is they are subjected to movement if the building shifts or the floor sags.

Concentrated load

Concentrated loads are heavy loads located at one point in the building. A steel beam resting on a masonry wall is an example of a concentrated load. Air conditioning units on a roof is another example. Concentrated loads must be anticipated and accommodated in the planning of the building. Many times, structural alterations that add an additional story or roof-mounted signs and HVAC units are not fully calculated into the loads that the existing structure is designed to carry, and the load becomes a nondesigned load. When this condition is present and fire further damages the structural members, this challenges the ability of the load-bearing members and can result in total structural collapse.

When a concentrated load is not a designed load, which is often the case with alterations involving heavy air conditioning units or other machinery, it increases the potential for a collapse danger. This necessitates an ongoing size-up of the fire building to monitor this type of force on load-bearing elements (fig. 2–18).

Fig. 2–18. Heavy air conditioning machinery often causes roof collapse when the roof supports are weakened by fire (courtesy FDNY).

How Loads Are Applied

Any structural member under a load must change shape or deform to some degree. Compressive loads will cause a shortening of the material caused by the force that tends to crunch or squeeze materials together. Tensile loads will cause the material to lengthen or stretch, and the force tends to pull material apart. Shear forces tend to rend or break material by causing its molecules to slide past each other.[26]

In bowstring truss construction, a load applied to the top chord of the truss creates compression, and this load is transferred to the bottom chord, which results in a pulling force known as tension. A similar effect occurs in beams and girders (girders support several beams or joists). The loads applied to the top of a beam cause compression, which pushes the beam to bend down (more in the center area of the beam than near the supporting ends). This causes the bottom area of the beam to have tension. Another similarity between beams and trusses is that the depth of a beam will determine its load-bearing capacity, and in a truss, the distance between the top chord and the bottom chord determines loads it can carry. Girders generally carry more weight and have greater depth than beams.

Distributed load

A distributed load is one that is passed over a large area and is carried by a larger number of supports that will take that load to ground. Consider a load at the mid-span of a joist. This load is split into two parts, with a proportional part going to each end of a beam supporting the joist. This in turn becomes a point load at the supporting four columns. They transfer the load to the earth. The load on each column in this case is only one-quarter of the initial load. This is one method of distributing a load.

Building loads from structural material weight, contents, and so forth are collected by structural members in the horizontal plane, transferred at where they connect to a beam, and then transferred to the columns, which bring the load to the foundation.[27]

Compression, tension, and shear are the basic forces that are created by a load. The force is dependent on the manner by which the load is applied to a material (fig. 2–19).

Fig. 2–19. Loads, in this case floor(s) above, create compression, tension, and shear forces on structural elements (courtesy FDNY).

Axial loading

An axial load is a force that passes through the center of the section under consideration. The load is evenly applied to the bearing structure. Generally, a structure will sustain its greatest loss when an axial loaded structural member fails.[28]

When a building is subjected to fire, loads that were once carried in an axial fashion now are altered. Shifting loads are one of the most common initial causes of structural collapse.

Eccentric loading

Eccentric loading is a force that is perpendicular to the plane of the section but does not pass through the center of the sections, thus bending the supporting members. The load is straight and true but it is concentrated to one side of the center of the supporting wall or column. Often, during the initial collapse event, loads that were once carried in axial fashion pass through the eccentric stage on their way to causing a collapse (fig. 2–20).

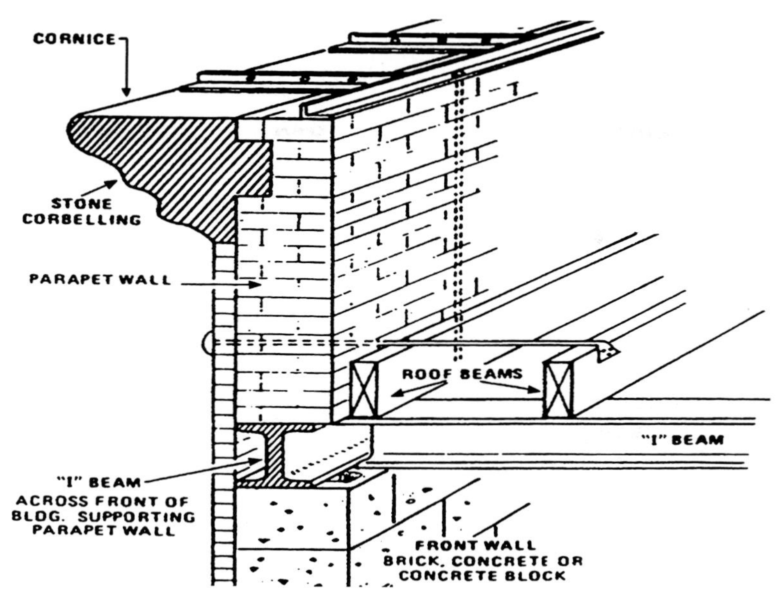

Fig. 2–20. A parapet wall I-beam creates an eccentric load force on the supporting wall.

An example of an eccentric loading is the parapet wall. A parapet wall is defined as a continuation of an exterior wall. This wall is eccentrically placed on the I-beam that rests on the front wall. To counter the eccentric force, tie rods are inserted into the wall and attached to an angle iron that is attached to a roof beam. This holds the parapet wall back into the building.

Fire Chronicle

A parapet wall of this type of design collapsed in 1981 during a fire on Dyckman Street in Manhattan. The building was a 125 × 100-ft, one-story commercial. Upon arrival, units were faced with heavy fire conditions in a row of stores. The chief made a survey of the building and discovered a bulge in the parapet wall. He immediately ordered an exterior attack. Firefighters in front of the building were removed and a collapse zone was established. Approximately 35 min later, the entire 125-ft-long parapet wall collapsed with debris covering the entire front sidewalk. The collapse was caused by the steel roof beams, which elongated as a result of being heated by the fire, pushing the parapet away from the building (fig. 2–21).[29]

Fig. 2–21. The parapet wall completely collapsed at this fire (courtesy FDNY).

* * * *

Steel beams begin to lose structural strength at 1,000°F. A 50-ft-long steel beam when heated evenly will expand up to 4 in.[30]

Torsion load

Torsion loads are forces that are offset from the shear center of the section. They are inclined toward twisting the supporting member. It is a type of load that causes objects to twist due to torque. This force can be demonstrated by twisting a piece of blackboard calk until it breaks. Torsional loads are basically twisting loads (fig. 2–22).

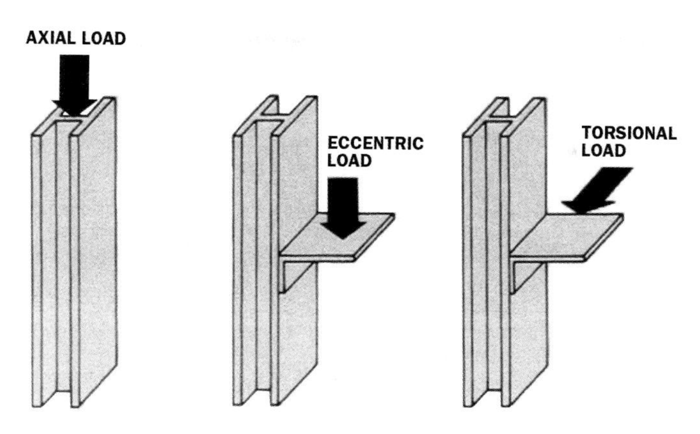

Fig. 2–22. Building load forces

Cantilever load

A cantilever is a beam supported on only one end. The beam carries the load to the support where it is resisted by moment and shear stress. Cantilever construction allows for overhanging structures without external bracing. Cantilevers can also be constructed with trusses or slabs.

Any structural member supporting a cantilever load, such as a marquee, should have the connections points checked for

- Deterioration of mortar in supporting masonry wall
- Rusting of supporting hardware
- Weight or excess live load such as snow or water
- Stress caused by sudden impact

Suspended load

Loads can be held in suspension with far more weight than can be carried in compression by columns. The load cannot be carried to ground in tension; it must be converted to a compressive load onto a beam or column. The points of connection are over the structure being supported. Thus another consideration of these load supports is that the upper area will be subjected to greater heat levels during a fire. Connection point failure is responsible for a large number of suspend load collapses.

On July 17, 1981, a collapse of a suspended walkway resulted in the deaths of 114 people in the Hyatt Regency Crown Center in Kansas City, Missouri. The lobby featured a multistory atrium crossed by suspended concrete walkways on the upper levels. When the walkways became crowded with people, this additional live load initiated a total failure of the structure. Connection point failure caused the fourth and second level walkways to collapse into the lobby below.[31]

It should be noted that steel cable or tie rods are most often used to suspend loads, and cold drawn steel fails at 800°F.[32]

Sheer stress

A shear stress is defined as a stress that is applied parallel or laterally to a face of a material, as opposed to a normal stress, which is applied perpendicularly. Weight is a force; hanging something from a wall creates a shear stress on the wall, since the weight of the object is acting parallel to the wall, as opposed to hanging something from the ceiling, which creates a tension stress. Sheer forces tend to cause materials to slide past one another.[33]

A steel fire escape has some of the same connection point weaknesses as overhanging signs on a building and is subject to several types of stress conditions. In addition to the sheer stresses, it also imposes an eccentric load on the exterior wall, because the load of the fire escape load does not pass through the center of the exterior wall. Fire escapes also suffer from weather damage, which further damages connection points and reduces their load-bearing capacity.

Fire Chronicle

As a captain in command of a ladder company operating at a loft building constructed of brick walls with wooden floors, I was ordered to search a floor area off the third-floor fire escape. There were several engine company members operating a hose line on the fire escape platform that I had to use. The officer of the engine company pleaded with me not to come out onto the fire escape platform, which had some sections slightly separating from the railing due to the weight of his members (I will never forget how wide his eyes were as he looked in my face). I notified the chief of the condition of the fire escape platform and suggested gaining entry to the area by going through the interior. The chief agreed to alter the operation. Company officers should be encouraged to apply risk management principles (which will be discussed in the next chapter) as conditions warrant and communicate safety concerns to the incident commander, rather than following orders at any cost.

* * * *

Building configuration

Square or rectangular buildings that have floor plans with symmetrically placed force-resisting elements tend to perform better when the structure is weakened than buildings composed of irregular shapes or those with large open foyers or lobbies that create a soft story condition (few main load-bearing members) (fig. 2–23).

Fig. 2–23. Square-shaped buildings tend to perform better when the structure is weakened than buildings composed of irregular shapes.

As with walls, the more connection points for distributing a load, in this case columns, the more resistance to collapse. Buildings with irregular shapes cannot distribute lateral forces evenly, which results in torsional load stresses that can increase damage at connection points in the building.[34]

Weight distribution

Buildings that are wide at their base and have most of their weight distributed to their lowest floors generally fare better when the structure is weakened. Within certain limits of the materials used, any type of load can be dealt with successfully. Axial loads are more easily dealt with by the building's designer than eccentric or torsional loads.

There is a delicate balance of forces in a building that can be undermined by nondesigned changes in loading. An example of this is when a failure of one part of a steel frame building or steel truss places nondesigned torsional, twisting stresses on other parts of the building, thus extending the area of collapse. It is this nondesigned shifting of the loading from axial to eccentric or to torsional that can weaken connection points, causing a collapse.

Connection Points

Connection points are often the initial areas of failure. There are several studies that support this. A study of fire-induced building collapse performed by NIST states: "The effects of elevated temperatures on the strength of connectors themselves and on their ductility, as well as how thermal expansion of adjacent heated members affects the stress redistribution in a floor and framing sub-assemblage through its connections, are important issues yet to be resolved. Connections are generally recognized as the critical link in the collapse vulnerability of all structural framing systems, whether or not fire is involved."[35]

The condition of the building's connection points is one of the major considerations in determining its stability. Every structural examination must include the areas where the structural members are connected to the rest of the building.

Fire Chronicle

I was the "all hands chief" (the chief officer responsible for supervising operations inside the building) at a fire in a church constructed of heavy timber trusses. The fire had started on the first-floor entrance area of the church and spread up into the choir loft above. In the choir loft area there was charring at the connection points where the heavy timber truss roof rafters were attached to the wood bearing walls. Smoke was venting through holes in the roof, which allowed a clear view of the rafters' connection points on the bearing wall. As this investigation was being performed, continuing extinguishing efforts were in progress, effectively controlling the fire that had spread up the walls and under the choir mezzanine.

Truss construction, combined with structural elements affected by the fire, and no life hazard—plenty of reasons to evacuate the building and let it burn, but I did not recommend that. Why?

The first determination to be made was the risk to operating members. The connection points did not show any deformity and were being protected by the streams from hose lines. All the visible connection bolts were intact. It appeared that the heavy timber rafters had sufficient mass remaining to sustain the roof loads and showed no distortion. There were no additional loads on the roof other than the original wooden sheathing and asphalt shingles. The water runoff from the hose line operations was draining out of the building and not adding to the existing loads. The bearing walls showed no deflection and did not suffer any significant damage. The building appeared stable.

The next determination was, did property of value exist? There was little damage to the structure or contents of the church with the exception of the front entrance and choir loft mezzanine area.

All this information was radioed to the incident commander outside, and he concurred with my assessment to continue extinguishing the fire from the interior.

I was fortunate in this situation in that I could scrutinize the vital connection points. This is often not the case under fire conditions where smoke obscures visibility. The main reason for citing this experience is that recommendations are not always to change strategies to exterior operations and let the building burn. There is always a struggle to weigh all the factors involved in making the call that will determine the outcome of the firefighting operations. As they say: "This is why the chief gets paid the big bucks."

It is critical to give accurate information to the incident commander. An officer's report must contain specific information regarding the condition of load-bearing structural elements and their connections to other structural members in order to be of use to the incident commander in determining the building's stability and its effect on firefighting strategies. Understanding what is occurring to the building's structure is essential if this is to be accomplished. This can only be achieved by learning how loads and stresses affect building stability. Once this knowledge is attained, the ability to effectively communicate the information becomes the key in providing the incident commander with an understanding of existing conditions and how to proceed. Common terminology gained from standardized training will help facilitate this understanding.

Final extinguishing efforts were achieved at this fire with the structure damage contained to the entrance area. There were no significant injuries to firefighting personnel. The church was repaired and opened shortly afterward to serve the community.

* * * *

Vacant buildings

In addition to duration and intensity of the fire and type of construction, also consider the prior condition of the building when assessing its stability. A vacant building should always be considered prone to collapse. There are many factors that may cause catastrophic failure: ongoing construction in the immediate area, water damage, rotting of wooden structural elements at connection points, and removal of structural members causing load shifts in the bearing walls. Proper time must be devoted to an exterior evaluation of these buildings when they adjoin a fire building (fig. 2–24). In a sense, a connection point can include an adjoining building when the buildings are attached.

Fig. 2–24. Proper time must be devoted to exterior evaluation of vacant buildings when they adjoin a fire building (courtesy FDNY).

Fire Chronicle

In New York City, a working fire was on the second floor, extending to the third floor of a six-story, 40 × 90-ft brick and wood tenement. Adjoining the fire building was a building with scaffolding on the front wall. An exterior survey of this building revealed a previous partial collapse with floor beams separated from the load-bearing walls. A safety zone was immediately established to include this adjoining building, and 10 minutes later all six floors of this building collapsed in a pancake fashion. There were no injuries to operating personnel.[36] Establish safety zones early to include weakened structures in the immediate area, and maintain them.

* * * *

Partial Collapse Dangers

A report of partial collapse must be investigated immediately to determine the extent of the stresses that are being put on the remaining load-bearing structural members in the area.

Whenever there is a partial collapse of a structure, there is a strong probability that the entire structure has been weakened by the event. The load-bearing forces that were supported by the structural elements that failed are now being distributed to other parts of the building. The existing structural loads are being increased and new loads applied in ways that these remaining structural members were not designed to accommodate. As stated earlier, the building configuration affects the way these loads are being distributed. An irregularly shaped building cannot distribute lateral forces evenly, resulting in torsional stress that increases the probability of additional collapses, which, in a chain reaction, further weakens the entire building.

Localized collapses require attention to the connection points, which must be checked for any distortions of load-bearing members caused by nondesigned loads. This is particularly significant when lightweight construction is involved where there is more dependence on other load-bearing members connected within the building to distribute loads and provide structural stability.[37] Also, the decrease of structural mass used in this type of construction further reduces the building's ability to withstand the effects of nondesigned load forces.

Initial withdrawal of operating personnel from the partial collapse zone should be considered until a determination is made regarding the stability of the building elements in the area. All members operating at the scene should be informed of the partial collapse zone and actions taken to ensure that only the personnel performing a structural inspection and/or life-saving operations are permitted in the affected area.

When a partial collapse involves upper areas in the roofs of churches or warehouses or other structures with very high ceilings, immediate withdrawal from the interior collapse areas of the building should be considered for several reasons. One reason for immediate withdrawal is if the fire is burning in an area that is difficult to extinguish due the limited reach of handheld hose streams. The time involved in replacing handheld lines with large-caliber streams required to effectively extinguish the remote fire further adds to the collapse dangers, as the fire is continually destroying the structural members while these operations are taking place.

The destructive force of the high-caliber stream's impact on the building combined with the dangers of serious injury from falling overhead debris and the likelihood of further collapse warrants removal of interior forces from the collapse zone. Prior to resuming interior operations in the collapse zone, an inspection must be made to assess the stability of the remaining structural elements. An incident commander should anticipate the difficulty of performing this inspection because of the height of the ceilings and adjust tactical objectives as necessary.

Large-Caliber Stream

Large-caliber stream operations deliver approximately 500 gallons of water per minute, which equals roughly 4,000 lb of water per minute of additional load into a fire-weakened structure.[38] Two tons of water speeding through a nozzle at 100 ft per second is a significant destructive force (fig. 2–25).

Where the stream is operating and for how long must be factored into the judgment regarding the continuing stability of the building. An article featured in *Fire Engineering* magazine highlighted a chief officer's knowledge of the effects that large-caliber streams can have on causing building collapse. "A fire in New York City burned out of control in a group of 126-year-old buildings where large-caliber streams were being used. The building department engineers were called in to assess the stability of the building at 1800 hours. Viewing the exterior walls, they observed that there were no cracks or bulges in the walls, the walls were not out of plumb, the window openings appeared straight, and the surveyor's transit (operated by FDNY personnel) detected no motion. The building was declared stable at this time. The chief in command of the fire was aware of the effects of the large-caliber streams in washing away the sand-lime mortar in the brick joints, and runoff water being absorbed by plaster and stock. The impact of the large-caliber stream on the buildings structural members was also a concern. The chief continued to enforce the safety zones and refused to allow anyone to enter the buildings. Not long after this decision, two of the buildings suddenly collapsed."[39]

Fig. 2–25. Large-caliber streams can deliver a load of 4,000 lb of water per minute (courtesy FDNY).

Fire Chronicle

I was a captain in command of the FAST company (Firefighter Assist Search Team) at this fire. The FAST team officer is positioned at the command post, so I had an opportunity to observe how this fire was run. The fire had gotten into the walls and was spreading through the interconnected buildings. Units would extinguish fire in a room, move on to another area, and the fire would flare up again where it had been extinguished. Hi-Ex foam was introduced into the cellar on two occasions with no positive effect. Ladder companies were heavily engaged in opening up the walls and ceilings to find active burning. Hose lines were repositioned continuously to battle fire breaking out in several areas at once. Interior operations were suspended while large-caliber streams knocked down heavy fire. When it looked like the fire was darkened down, the handheld hose lines went back in, only to find fire breaking out in the same places where it had been extinguished before. It was a hard fought, frustrating, debilitating process. In a word, "ugly." And it went on for hours. When the building engineers concluded that the buildings were stable and safe to enter, I thought the chief would allow members back in the building to fight the fire once again. There were large areas of the buildings that had not been affected by the fire and all the buildings looked relatively stable. Many of the storeowners on the scene were imploring the chief to let them back into their stores. Knowing the damage caused by large-caliber streams, he refused to allow anyone into the buildings. Very few expected the sudden, extensive collapse that occurred. I know I didn't expect it and the storeowners who were clamoring to get back in the buildings certainly didn't. This chief relied on his experience and his judgment was correct. A lesson I took notes on.

Another thing I recognized at this fire was that the fire chief owns the fire building. No one—not the building owner, insurance company, building inspector, police department—no one, can enter that building until the chief says it is safe to do so.

* * * *

Firefighters actually witness more building collapses than building department personnel or civil engineers, and have more firsthand experiences. Consult with knowledgeable professionals whenever you can, but do not disregard your understanding of what has occurred to affect the stability of the fire building.

Incident commanders should frequently perform a size-up to access the changes that have occurred. Be aware of the fact that the firefighting efforts are continually adding water loads and the building is constantly being weakened by fire.

Progressive Collapse

A collapse in a progressive manner can lead to a total failure of a building's structural supports. Highrise building owners have realized the potential for large numbers of deaths and property loss resulting from a collapse of a highrise building. This has led to an interest in evaluating the potential for progressive structural failure of existing buildings and design features for new construction to resist progressive collapse. Structural failure caused by earthquakes has also led to studies on the characteristics of progressive collapse. Fires and earthquakes have many similar destructive effects on a building's structure and the lessons learned from one event can often be applied to the other.

Progressive collapse is defined as the spread of an initial failure from element to element, eventually resulting in the collapse of an entire structure or a major section of it. This can occur when a local failure of a primary structural member, such as a column, causes adjoining members to fail which then continues through the structure unless adjoining structural members arrest further progression of failure (fig. 2–26).[40]

Fig. 2–26. Progressive collapse is defined as the spread of an initial failure from element to element, eventually resulting in the collapse of an entire structure or a major section (courtesy FDNY).

Investigations into the causes of progressive collapse have been performed which have resulted in proposed solutions to prevent progressive collapse. Dr. W. Gene Corley, who worked as a structural engineer for over 35 years, gave this testimony:

> Redundancy is a key design feature for the prevention of progressive collapse. There should be no single critical element whose failure would start a chain reaction of successive failures that would take down a building. Each critical element should have one or more redundant counterparts that can take over the critical load in case the first should fail.[41]

There are limits in assessing the vulnerability of new and existing buildings to progressive collapse. An analysis of progressive collapse was performed by H.S. Lew to check the capability of the structural system to resist removal of a specific critical structural member. This study takes into account the difficulty of determining the patterns of individual factors:

> The potential progressive collapse analysis entails evaluating a structure for its vulnerability to the development of a partial or a total collapse of the structure initiated by an event that causes local damage. At the present time, there are no accepted simple analytical tools that design professionals could use to analyze progressive collapse potential of low- and mid-rise buildings.[42]

When it comes to finding reliable information on the collapse potential of a burning building, you keep coming upon the same situation: "Until more data becomes available the conclusions are limited." The lack of structural behavior data associated with a progressive collapse caused by fire damage inhibits the ability to provide comprehensive information for fire ground commanders. However, an awareness of this phenomenon is necessary due to its catastrophic consequences.

The information from fire department members who witness large numbers of partial and complete building collapse becomes even more valuable due to the lack of accurate scientific analytical methods. A fire officer must place reliance on past experience and known structural failure characteristics to determine strategic decisions.

Protecting Operating Personnel

There are several ways to protect personnel from the dangers of falling debris from a building collapse.

1. Know the way a wall can collapse and how far the wall may fall. For example, a 90-degree angle collapse, an inward/outward collapse, a lean-to collapse, tent collapse, pancake collapse, and V-shaped collapse.
2. Establish a collapse zone early and maintain it.
3. Use the reach of hose streams to keep members out of the collapse zone.
4. Flank a dangerous wall if necessary.
5. Report collapse hazards to the incident commander and make sure all operating members on the scene are aware of the dangers.

Elevated large-caliber stream operations

An increasing number of firefighters have been killed or seriously injured operating large-caliber streams in tower ladder buckets. To protect the members operating the large-caliber streams, keep the bucket a horizontal distance from the building wall at least equal to the height above the bucket. That is considered the collapse zone for members in the bucket.

Large-caliber streams may knock debris off the building, which requires expanding the collapse zone on the ground as well (fig. 2–27).

Determining factors in maintaining the interior fire attack

A major factor in continuing the interior attack is whether or not the fire involves contents or structure. Generally, if only the contents are involved and there are no indications of significant structural weakness, then, with the proper personnel, the interior fire attack is indicated to save lives and property. The combustible contents are the primary fuel for most structure fires. There has been a large increase in plastic products in all types of occupancies, creating fire loads that increase the temperatures, toxicity, and spread of fire. These dangers of plastic combustion inside a burning building with lightweight construction create a significantly high risk to firefighters who

Fig. 2–27. Consider the collapse zone for elevated platform operations (courtesy FDNY).

are rapidly penetrating into the structure wearing protective equipment necessary to operate in this environment. New designs for lightweight air packs and encapsulated bunker gear allow firefighters to advance quicker and deeper into a burning structure than they did in the past. This change in tactical operations increases the importance of making rapid risk management decisions by the incident commander.

When did the fire progress from a content fire to a structural fire? This is a prime consideration and is often very difficult to calculate. Did a particular condition develop while units were on the scene or did it exist prior to arrival?

To calculate how long the load-bearing elements have been subjected to fire damage prior to arrival, consider when the fire occurred. If the fire began in the middle of the night, add a considerable amount of time. For a fire that started during the daytime, a shorter time estimate would be appropriate. If there are witnesses available, ask them if they have any information regarding the time they noticed the fire. Some type of calculation regarding when the structure became involved in fire must be determined. It often comes down to guesstimating a degree of probability the best you can. Plan for the worst, and start the clock for structural damage when the fire was first reported if there is no other information available.

Tips from the Pros

During the time I was teaching the chief officers of the FDNY courses on building collapse, I always asked if there were any other bits of information that they could add. The following information was gathered from their responses:

- If you are anticipating the collapse of the fire building, take into the account the effect this building collapse will have on adjoining buildings, and include them in the collapse zone if required.
- When taking a defensive position by flanking the building in corner areas, keep in mind that these areas are less dangerous than other areas but are not completely safe.
    - Use maximum reach of the hose stream to further increase the safety factor.
    - When the collapse zone has been established, as soon as possible, change handheld hose lines to large-caliber stream appliances.
    - Firefighters have a tendency to creep back into the collapse area when operating hand lines.
    - A collapse zone has a tendency to shrink, so start out with the largest possible area.
    - Secondary collapses often occur after the primary collapse. Particular attention must focus on any unsupported walls.
    - Do not underestimate the knowledge you have gained observing past building collapses caused by the effects of fire. But keep in mind that due to the increase in buildings constructed of different combinations of new materials and techniques, this limits some of the value of past experience in determining a building's collapse potential.
    - Supplement your awareness by consulting with knowledgeable people or experts in collapse whenever possible.

- Try to find someone familiar with the fire building for additional information about any unique structural characteristics.
- Have a surveyor's transit set up to monitor if the building is moving. Constant communication with personnel using this equipment must be maintained.
- A change in the original occupancy of the building usually will be accompanied with major renovation work to the structure.
- Alterations have often been found to be a major factor in setting off the collapse of a burning building. Anticipate additional structural hazards and weaknesses whenever alterations have been performed to a fire building.
- Encourage firefighters of all ranks to bring collapse concerns to the attention of the incident commander. A member's rank should not prevent him or her from communicating concerns.
- After the fire has been extinguished, if it is necessary to reenter the building, limit the load applied by keeping the absolute minimum number of members in the fire area or on stairs.
- As soon as possible, have the utilities to the building shut down to reduce electrical hazards and additional water loads from leaking pipes.
- And lastly, stick to the procedures, they work!

I sincerely thank all those chief officers for their insights and contributions to the safety of firefighters.

References

1. Federal Emergency Management Agency (FEMA). *Earthquake Hazard Mitigation Handbook*, "C: An Introduction to Structural Concepts in Seismic Upgrade Design." Retrieved from conservationtech.com, November 2011.
2. Brannigan, Francis. *Building Construction for the Fire Service, Third Edition*, Quincy, MA: National Fire Protection Association, 1992, 202.
3. Brannigan, 162.

4. Dunn, Vincent. *Collapse of Burning Buildings: A Guide to Fireground Safety*, Tulsa: Fire Engineering, 1998.
5. O'Connell, John P. *Emergency Rescue Shoring Techniques*, Tulsa: PennWell Corporation, 2005.
6. *Polystyrene Forms User Manual*, American Polysteel Forms, Third Edition, 1995, Albuquerque: Berrenberg Enterprises.
7. Dunn, Vincent. *Safety and Survival on the Fireground*, Saddle Brook, NJ: Fire Engineering Books & Videos, 1992.
8. FDNY First Line Supervisors Training Program, Building Construction Course.
9. Brannigan, 64.
10. Norman, John. *Fire Officer's Handbook of Tactics, Third Edition*, Tulsa: PennWell Corporation, 2005, 426.
11. Brannigan, 171.
12. Ibid., 166.
13. FDNY First Line Supervisors Training Program, Building Construction Course.
14. El Shayeb, Mohamed, et al. "Utilization of Numerical Techniques to Predict the Thermal Behavior of Wood Column Subjected to Fire, Part C: Sensitivity Analysis," Paper presented at the 6th International Conference on Fracture and Strength of Solids, (FEOFS) 2005 Bali, April 2005. Universiti Tenaga Nasional.
15. Anderson, L.O. *Wood-Frame House Construction*. New York and Hong Kong: Books for Business, 2002.
16. Spadafora, Ronald. "The Fire Service and Green Building Construction: An Overview," *Fire Engineering*. Retrieved from www.fireengineering.com, November 2011.
17. Ryan, James V., and Edward Bender. "Fire Endurance of Open-Web Steel-Joist Floors with Concrete Slabs and Gypsum Ceilings," *National Bureau of Standards, Building Materials and Structures Report*. August 1954.
18. Dunn. *Collapse of Burning Buildings*.
19. Quintiere, J.G. "Fire Investigation, An Analysis of the Waldbaum Fire, Brooklyn, New York, August 3, 1978," *NISTIR 6030*, National Institute of Standards and Technology (NIST), June 1997.
20. FDNY Fire Fighting Procedures, Truss Roofs.
21. Montagna, Frank. "Hazards of Membrane roof Fires," *Fire Engineering*, July 1998, 61.
22. NYC 2008 Building Code Section 1602.
23. NYC 2008 Building Code Chapter 16.

24. Iwankiw, Nestor, Craig Beyler, and Jess Beitel. "Testing Needs for Advancement of Structural Fire Engineering," *Proceedings of the Fifth International Conference on Structures on Fire, 2008*, 334–343.
25. Brannigan, 30.
26. Ibid. 45.
27. Hammond, David J., S.E., and Paul DeCicco, P.E. *NFPA Fire Protection Handbook*, "Section 12.11." National Fire Protection Association, 2003.
28. Brannigan, 27.
29. Miale, Frank. "What Brought Down the Parapet Wall This Time?" *With New York Firefighters* (*WNYF*), Fourth issue 1998.
30. Brannigan, 33.
31. Brunet, Edward J., Jr., P.E., "Engineering Ethics Lessons Learned: Kansas City Hyatt Walkway Collapse." Retrieved from www.pdhengineer.com, November 2011.
32. Brannigan, 33.
33. Ibid., 45.
34. Taranath, Bungale S. *Wind and Earthquake Resistant Buildings: Structural Analysis and Design*. New York: Marcel Dekker, 2005.
35. Beitel, Jesse and Nestor Iwankiw, "Analysis of Needs and Existing Capabilities for Full-Scale Fire-Resistance Testing." National Institute of Standards and Technology (NIST) *GCR 02-843-1*, September 2008.
36. FDNY Safety Bulletin 57, Case Study 5 Collapse—Vacant Building.
37. Brannigan. 523.
38. Dunn, Vincent. "Large-Caliber Stream Safety," *With New York Firefighters* (*WNYF*), third issue 1998.
39. Brannigan, Francis. "The Building Department Said the Walls Were OK, But ..." *Fire Engineering*, May 1998.
40. David N. Bilow, P.E., S.E. and Mahmoud Kamara, PhD. "Progressive Collapse Analysis and Design for New Federal Office Buildings and Major Modernization Projects," Portland Cement Association, September 2003.
41. Corley, W. Gene, P.E, S.E. Testimony given before the Public Buildings and Economic Development Subcommittee, House Committee on Transportation & Infrastructure, United States House of Representatives, June 4, 1998.
42. Lew, H.S., "Analysis Procedures for Progressive Collapse of Buildings," Building and Fire Research Laboratory, NIST, Gaithersburg, MD.

Risk Management for the Fire Service

3

Knowledge of the effect fire has on load-bearing structural members, how load forces affect the structure, and how different materials are weakened by fire is essential for a fire officer. Knowing where and what to look for regarding structural weaknesses will provide the basis for competent decision making. But this by itself is not enough.

A fire officer commanding firefighters at a building fire must know how to apply this information in an appropriate manner. An understanding of risk management principles can provide solutions for the decision-making process that must occur.

At different times various management theories have been used by the fire service; for example, in the 1960s personnel management techniques were introduced, and in the 1970s management by objectives was studied. The methodology used most often today, known as risk management, is the discipline of identifying, monitoring, limiting, and mitigating risks. Risk management is used to understand the reality of managing on-scene firefighting operations.

It is crucial for supervisors of firefighting personnel to understand the concepts of risk management in order to identify and control the risk firefighter's face on the fireground. When the International Association of Firefighters investigates a firefighter fatality, a fundamental question that must always be answered is, "Was it appropriate for firefighters to be in the building?"[1]

No active firefighting operations are ever risk free (fig. 3–1).

Firefighters expect and accept a degree of risk during routine firefighting and emergency operations. Fireground risk management is a tool for determining which risks are acceptable. The basic concepts are

- Do not risk a lot for a little.
- Consider the odds. How severe are the consequences if things go wrong? When considering the odds, consider the worst possible scenario.

Fig. 3–1. Practically all operations on the fireground involve the risk of physical injury.

Fireground Guidelines

Every life and death decision made on the fireground can be made according to two fireground guidelines:

- Protection of life, both civilian and firefighter
- Protection of property

The basic principles of the risk/benefit analysis are made upon these two guidelines. When firefighters enter a burning building and quickly extinguish a fire with the initial attack hoseline, this action is the most effective method in accomplishing the first and second fireground goals and satisfying the priorities of risk taking (fig. 3–2).

If a fire is allowed to grow and extend, this increases the potential for firefighters to be killed or injured (fig. 3–3).

Fig. 3–2. When firefighters rapidly extinguish a fire with the initial attack hoseline, this complies with the goals of risk-taking.

Fig. 3–3. A fire extending inside the structure increases the risk.

Each action on the fireground carries with it a benefit and a risk. The acceptable level of risk is directly related to the potential to save life or property. Where there is no potential to save lives, the risk to fire department personnel must be evaluated in proportion to the ability to save property of value (fig. 3–4).

Fig. 3–4. The risk to fire department personnel must be evaluated in proportion to the ability to save valuable property.

Acceptable risks are those where the benefit is of more importance or value than the negative possibilities posed by the situation.

Emergency responders may risk their lives in a calculated manner to save a life and place themselves in a situation with moderate risk to save property.

The FDNY's safety bulletins provide guidelines regarding risk management:

> The risk to be assumed by our members must be in proportion to the expected gain. Thus search tactics at an occupied tenement should be substantially different from those in a vacant building. While there may be a suspected life hazard at a vacant building, the only known life hazard at a vacant building is that of the members. Chief and company officers must be cognizant of the risk members are exposed to when operating at vacant building fires.[2]

See figure 3–5.

Fig. 3–5. Identify and communicate hazards.

Routine operations to protect property will involve risks to personnel. Whenever possible, measures should be taken to reduce or avoid risks associated with these activities.

Where there is no ability to save lives or property, there is no justification to expose firefighters to any avoidable risk. For these situations, the appropriate tactics should be exterior or defensive fire-suppression operations limiting the spread of fire from the original structure or area.

Principles of Risk Management

Incident commanders are responsible for constantly evaluating the risk to members with an understanding of the reasons for their actions in each situation.

Risk management principles provide a practical method to judge and control the necessary risks required to accomplish firefighting goals.

The four "principles of risk management" are

- Risk identification
- Risk evaluation
- Risk prioritization
- Risk control

Using these principles of risk management will help determine what actions to take when attempting to define fireground risks and how to control the risks being taken.[3]

Risk identification

Some things will be easily identified as a risk; for example, a warning sign indicating a specific hazard for high electrical voltage. After identifying a particular hazard, the incident commander and all firefighters on the scene must be made aware of the hazard. *Identify* and *communicate* what the known hazard is immediately (fig. 3–6).

Fig. 3–6. Overhauling after the fire is out can be considered a high frequency/low severity event

When encountering a unique event or a large fire area upon arrival, it can be very difficult to rapidly identify the various risks accurately.

To effectively identify fireground risk, use all the resources that are available to you, including

- Accident/injury statistics
- Fire case studies
- Your experience and that of others

There will be different risk for different situations; gas leaks, expanding fire conditions, building collapse. When identifying the risk, formulate a safety plan for the worst-case scenario; anything less becomes easier to handle. Should risk identification be performed in a rapid manner?

Fire Chronicle

An alarm was received for a building fire in the middle of the afternoon in Brooklyn (not an unusual occurrence). The location of the fire was close to the battalion headquarters where I was stationed. I arrived on the scene soon after the first-due units. There was fire showing out of the first-floor windows of a two-story, wood frame, vacant, detached private dwelling. Members were about to enter the first floor with a hose line and were positioning ladders to search the second floor. I gave a clear message over the handie-talkie radio: "No one is to enter this building. We are going to use an exterior attack on the fire." It didn't take long to establish a collapse zone, set up large-caliber streams from portable master-stream appliances, and perform an exterior operation that eventually extinguished the fire.

It is important to recognize that you cannot immediately stop all operations on the fireground once there is a commitment to a particular strategy. The full withdrawal of members operating within a structure requires certain actions; for example, hose lines protecting egress must remain in position until members can get out and a roll call assures that everyone is accounted for. It takes time to communicate and receive information from units performing different tasks at different locations. It is like a military procedure. You do not immediately turn an army around on a dime when they are heavily engaged in battle. The same applies to a large commitment of firefighting forces engaged in an interior attack on a fire. The complexity and efforts required to change from an interior attack to exterior attack is often underestimated. These must be factored into a chief officer's strategy.

I was fortunate at this fire because I was able to identify the risk quickly.

This was a vacant building with no apparent life hazard or property of value, so there was no reason to risk operating personnel. I changed the tactics from the standard operating procedures of an aggressive interior attack to a defensive exterior attack *before* members took their positions for interior operations. Perform risk identification as soon as possible to help determine appropriate tactics.

* * * *

Risk evaluation

Once the risks are identified, the second principle of fireground risk management is applied: risk evaluation.

To evaluate a risk take into account the frequency and severity factors of the incident. What is the frequency of the threat? Some risks are present at every incident and others are faced maybe once in three years. Severity is determined by judging the negative consequences to operating personnel.

Judging the factors of frequency and severity together will help establish the priority of actions that must be monitored. These two factors are often grouped in the following manner to assist in understanding how to evaluate an incident:

1. High frequency/high severity
2. High frequency/low severity (fig. 3–7)
3. Low frequency/high severity (fig. 3–8)
4. Low frequency/low severity

Fig. 3–7. Low frequency/high severity (courtesy FDNY)

Fig. 3–8. A fireground often has multiple dangerous activities occurring simultaneously.

High frequency/high severity factors occur at every significant fire and often involve serious burns and deadly smoke inhalation. High frequency/low severity events are sprain and strain type injuries that result from routine operations such as overhauling after the fire is extinguished. Low frequency/high severity would be the dangers firefighters encounter at a building collapse. Low frequency/low severity events are minor injuries that do not occur often. Analyzing these factors helps us to organize our actions.

There are many operations being performed on the fireground in different areas; searching for victims on different floors, ventilating the roof of the fire building, examining adjoining buildings, and so on. An incident commander cannot be everywhere at once. We evaluate the risk in order to prioritize which risk to monitor and control at the scene.

Generally, the high severity situations should be addressed first and the low frequency/low severity issues are addressed after assessing the potential for all other possible events.

For each particular event, use your personal experience to look ahead from the initial phase to what might develop as the situation progresses.

Risk prioritization

This can be a difficult task. At an emergency where multiple dangerous activities are taking place, which one should you turn your attention to first? How do you decide which one is next? Establishing priorities is different for every event and is based on several factors (fig. 3–9):

- Availability of resources
- Ease/difficulty of addressing each risk
- Time involved

What is the availability of resources on scene to manage a rescue operation? How difficult is it to manage events connected with the risk? Are there simple or complex solutions that might require extensive personnel or special equipment? What are the consequences if things go wrong? What if the rescue crew needs rescuing? How long will it take to overcome a problem? How quickly can you get help to the scene?

Fig. 3–9. Unprotected metal C joists in a fire building would be considered a priority risk (courtesy FDNY).

Personal judgment and experience play a major part in this activity. Keep it simple. The process begins with any risk that has a high likelihood of happening. Deliberate on the severity potential for damage and injury. This should not take long. You will be able to calculate the frequency and severity factors very rapidly, and they will help you decide which area can produce the most negative consequences. This will help determine which operations should be stopped or allowed to continue. When trying to decide what fireground actions need to be performed, trust your common sense. Your knowledge of what can occur and your training will support you in making the right decision.

Fire Chronicle

WTC FAST team activation. Initially after 9/11, daily operations at the site of the WTC required one team of FDNY members kept at each command post to act as the FAST (firefighter assist and search) team. The sole duty of this team was to respond to any situation where a firefighter or rescue worker was injured or needed help. They were equipped with high-powered binoculars to survey the zone, looking for any signs of danger. One night the captain of the FAST team told me he was watching three firefighters who had gone over the top of a very large rubble pile and had not come back in sight. The FAST unit was immediately sent to the area where they were last observed. I had all chief officers suspend victim-searching operations and conduct a roll call of all members. The role call was completed with all members accounted for. The FAST team found the members who disappeared over the pile; they were from an engine company who had been special called to the site to provide a pumper for water fire extinguishment. They had decided to go off on their own to check out the site when they wandered out of our zone.

I gathered them in front of my command post, telling them that because of their irresponsible actions I had to stop all activity until they were accounted for. I sent them over to the staff commander with instructions that I wanted them removed from the site. The incident demanded these priority actions:

1. Immediate response by the FAST team to search for possible trapped rescuers (high severity incident).

2. Suspending all operations to gain control of the area and reduce any further risk being taken by operating members. There was only one FAST team available and if another rescue situation developed, I would not have the resources to respond to the incident (evaluating and controlling risks).

3. Accounting for all personnel at the scene. Again part of the prioritizing process. This was required, but the response of the FAST team was the first action that needed to be taken.

Identifying the risk and evaluating the risk at the scene was followed by prioritizing and risk control. The first step is to identify the risk. That identification resulted in having a rapid response rescue team at each command post available to perform immediate actions to save rescuers. The order of risk evaluation and prioritizing can shift as necessary. The process always ends with taking actions to control the risk.

* * * *

Risk control

Once the priorities have been determined, the next course of action is risk control. Are the operations being performed appropriate and if not, what needs to be changed? Risk control is where the rubber meets the road. Performing all the preliminary steps and not acting to control the risk is futile. All fire officers have a responsibility to play a major role in determining if the benefits outweigh the risk.

Fire Chronicle

I was a battalion chief working in Brooklyn when I responded to a 3 a.m. alarm for a building fire with people trapped. Upon arrival we had heavy fire on the first floor of a three-story residential building, approximately 25 × 60 ft. My initial size-up indicated that the building was recently built, but I did not know what types of materials or methods were used in its construction. I suspected the worst: lightweight wooden trusses or metal C joists.

There was a woman and a child at the third-floor front window who needed to be removed. The building had only one small interior stairway that couldn't be used due to heavy smoke rising up from the fire in the first floor. I was receiving reports that there were people trapped on the upper floors in the rear. The first-due engine had a line stretched into the first floor. I followed the line in to try and determine if the fire was affecting the structure. I found the hose line was operating on fire in the rear, but heavy smoke obscured visibility.

I called the engine officer on the handie-talkie and asked if the fire was in the contents or the structure. He gave me the short answer that you would expect; "I don't know. The smoke is too heavy to see anything."

I was confident this company would aggressively attack the fire, and that the sheetrock should be protecting the load-bearing structural members at this point. What I didn't know was how the building was constructed. I needed to find this out.

I returned to the front of the building where portable ladders were now in position on the second and third floors, with firefighters going up to remove the people at the windows.

Everyone was very busy. I saw the second due ladder company chauffeur, who had a hook in his hand. I told him to open up the ceiling near the front door. He poked the ceiling a couple of times and the sheetrock came down. By now heavy black smoke was rolling out of the door and I couldn't see into the hole.

I took him over to a similar attached building where the front door was open and had him open up the ceiling. Now I could see how the building was made. It had metal C joists connected to cinder block, load-bearing walls.

C joists have no fire protection rating. If the sheetrock came off and left these joists exposed to fire, some degree of collapse would be imminent.

I stepped back and again performed a size-up from the front of the building. There were searches in progress and victims being removed from the upper floors. At this time in the morning, and with fire in a building I had never seen before, I was contemplating several possible actions. There was a large commitment of forces in this building that had known collapse potential with heavy fire on the first floor. What do I do? Do I withdraw members from the building or stay with the interior attack?

I radioed to the engine company, asking how much fire they had and if they were making progress. They responded that there were two rooms of fire left and that they were moving in on it. I knew this company and felt they would make short work of the fire. I then contacted the ladder companies on the upper floors to find out if there was any fire extension. They told me there was no extension at this time. I also received progress reports on the victims being removed and the continuing searches.

To order a complete evacuation at this point would endanger the members on the upper floors and possibly overlook fire victims who were not yet discovered. I decided to continue with the strategy of an aggressive interior attack.

After a short period of time (that felt so much longer), the fire was extinguished. It was primarily a contents fire with little structural damage to the building; the sheetrock held up. A typical residential fire.

* * * *

Why did I continue with a strategy that was risking serious injury to operating members?

If I had received reports of fire extension on the upper floors or of a delay in extinguishing the fire, or indications that the fire had penetrated the sheetrock and was involving structural members, I would have ordered the ladder companies to stop all searches immediately and withdraw from the building. I would have ordered the engine company to start backing out of the fire building but to keep their line controlling the fire until all members safely evacuated the building.

I knew from experience that sheetrock generally will not fail and can effectively withstand the damaging effects from fire. But you never know what kind of shape a building is in. Were there holes in the walls or ceilings? Were there any alterations in progress?

I applied my risk assessment; saving known victims from the windows and apartments above against the possibility of structural failure. I was risking a lot to save a lot. I identified the risk: lightweight structural members that might be affected by fire, which could cause a collapse of the building.

I evaluated the risk. Attacking and controlling the fire was primary; no rescues could take place without it.

I established priorities. Monitor the building for any possible warning signs that structural elements were being weakened by the fire.

I controlled the risk by maintaining the effective extinguishing efforts and rapidly removing victims.

It was good judgment, but if things went wrong and the building collapsed, I would hope I could still believe that. These are the types of decisions that a chief officer must make and they are not easy ones.

An important factor at this fire was that I had adequate manpower to address the life hazard and extinguishment efforts. Another factor was having adequate manpower to protect operating members. Committing members inside the fire building requires a high degree of certainty of the perceived benefits and the ability to accomplish those goals.

The following week, I was teaching a course on building collapse to newly promoted chief officers, impressing on them the collapse dangers of new construction. I told the chiefs about this fire to express some of the complexity and value judgments that must be made and how using risk management can help in the process.

Many of them commented that if the fire involves a building constructed of wooden trusses or C joists, we *must* fight it from the outside because these buildings are unsafe for us to operate in if there is a fire of any consequence. There was a lot of support for a change in strategy: from the time honored solution of the aggressive interior attack to save life and property, to an exterior operation to protect firefighter's lives in buildings with no fire protection.[4]

Rescue efforts can be justified in a building that is structurally sound but may fail from fire damage as operations continue. Once the buildings occupants have been evacuated, the building should be evaluated and tactics changed as appropriate to a new risk analysis of expected goals.

Alter, Suspend, or Terminate Operations

To lessen or remove hazards at the scene, it is sometimes necessary to change tactics. There are three ways to accomplish this: *alter*, *suspend*, or *terminate the operation*.[5]

Alter operations

Altering actions requires some modification of the procedures being performed to accomplish a task. For example; a company is operating a handheld hose line on the sidewalk that is inside the collapse zone to extinguish fire through the first floor front windows. By having the company move out of the collapse zone and operate their hose line from a safe vantage point, the risk is controlled by altering the procedure (fig. 3–10).

Fig. 3–10. Altering operations (courtesy FDNY)

Suspend actions

To *suspend* actions involves delaying procedures that are occurring. For example, a ladder company wants to proceed above the fire floor to perform a search for victims while the engine company operating the hose line on the fire floor is having difficulty containing the fire. Suspend the movement of the ladder company members to the floor above the fire until progress is made in controlling the fire on the fire floor. Once the engine company is moving its hose line in on the fire, knocking it down, the ladder company is then allowed to proceed to the floor above.

Fire Chronicle

During the first few days after 9/11, the weather was unusually mild for this time of year and didn't hamper the rescue efforts. This changed late one night some days into the rescue efforts when heavy rains with strong winds blew down onto the site. Large pieces of cracked glass began to fall from the World Financial Center buildings, which lined the rear of the "Liberty" command sector. The heavy glass scaled down into the site, smashing around the search teams and my Liberty command post. Anyone struck by this glass would suffer serious injuries. The heavy rain in the pitch black night was restricting vision and also causing loose debris to slide, making the unstable conditions in the pit very unpredictable. I decided to suspend operations by pulling all the teams back off the pile and moving to an area of safety. I called the on-duty staff officer and told him I was pulling my people back. He said that was my decision and shortly after, all sectors stopped work as well. That was the first time since the collapse that the intense, all-out activity at the site stopped. The rain and wind ended about two hours later and we were able to go back to work. During the pause I took the opportunity to have chief officers conduct a roll call to update the active personnel on the scene and provide accountability on who went back in, where they went to work, and who was supervising them. Once they were out on the pile again, the intense pace of searching for victims immediately started right up.

* * * *

The risk management process was applied as follows:

Risk evaluation. In this situation there was a need to risk a lot, but even these dire situations have to be evaluated for appropriate risk versus rewards. The "pile" was even more dangerous during this rainstorm. This increased the probability of having to rescue the rescuers and required a change in tactics. The risk was controlled by *suspending operations* until the weather conditions became stable. Once the rain and heavy winds stopped and conditions returned to where the searches could progress in productive ways, we resumed operations.

Accountability. Accountability was a major challenge during the initial rescue phases of operations at the WTC site. There were volunteers, retired FDNY members, personnel from different agencies, all operating in my "Liberty" command area of responsibility.

Some basic principles of accountability are:

- The incident commander must maintain an awareness of the activities and location of all units in his or her area.
- Company officers are obligated to know the location and condition of their crew members.
- All operating members are responsible for actively participating in the accountability process by remaining under the supervision of their assigned officer.

At any opportunity afforded, I performed accountability procedures by checking names against company/crew lists when they reported in or when rotated for relief, and having the supervisor responsible for tracking multiple crews perform periodical personal contact with company officers confirming status of operations and personnel. A chief officer was assigned the duty of maintaining an up-to-date list of all members and the supervisors they reported to. This was the only task the chief was assigned, and with over 150 personnel to keep track of, it required this level of attention to be certain of the accuracy of who was where.

These were the actions required to control risk and were part of my risk management plan. A personal accountability system must be used at all incidents, no matter how big or small.

Having the personnel available to perform this critical function places another burden on fire department leaders. This burden is to educate public officials why additional personnel is required to facilitate the ongoing tracking and accountability of assigned companies.

An incident commander is often working with companies for the first time at multiple alarm/mutual aid incidents. He or she must be prepared with reliable methods to account for personnel, and must take measures to ensure this goal is accomplished.

Terminate operations

Terminating an operation is a last resort and only performed if there is a significant life hazard involved. Terminating an operation requires all actions being performed to come to an end. Efforts being taken to complete a particular task cease completely. An example of this would be fire has extended behind members, cutting them off from their means of egress. They must be removed immediately from the fire area (fig. 3–11). Any procedures these members were performing are terminated and all efforts are now directed toward their safe removal from the untenable area.

Fig. 3–11. Terminating operations (courtesy FDNY)

Altering, suspending, or terminating operations are often necessary to effectively control risks.

Risk versus Reward Analysis

The risk/benefit analysis should be continually performed.

During the 9/11 WTC disaster, the principles of risk management were used to determine search procedures. Initially, when there was a high possibility of rescuing live victims, members placed themselves at great personal risk by operating in extremely unstable building collapse debris. Under normal circumstances, baskets of medals would have been awarded for the actions being taken.

Sometime after the second week, site operation procedures changed from a rescue mode to a recovery mode. The staff officers determined that no one could still be alive without having water for that period of time. High-risk rescue type searches were generally terminated unless there was some significant indication of a trapped victim that would justify a high-risk rescue effort.

To stop searching for live victims was a difficult decision to make. The principles of risk management made the need for this change in operations obvious to all and supported rapid acceptance of the decision. All actions taken must be communicated to the incident commander as soon as possible. These actions will affect the plans the incident commander has put in place and can influence the outcome of the event. The golden rule is, "Keep the boss informed." When terminating an operation, it is mandatory to immediately contact the incident commander and communicate what has occurred (fig. 3–12).

Frequently, the initial tactics employed at the scene of an emergency involve rescue operations. Officers must continually evaluate the status of potential victims and determine whether or not the procedures should change. Studies have shown that during confined-space rescue attempts, often the rescuers become fatalities attempting to recover dead bodies.[6] All officers must evaluate the situation continuously and ensure that the incident commander is aware of the risks taking place during an operation and the consequences if something goes wrong. Vigilant attention, constantly looking for changes occurring, is a trait of an effective officer. A risk/benefit analysis must be performed regarding the actions being taken to rescue potential victims or to save property of value, versus the safety of operating personnel. The success of tactical objectives must be balanced against the dangers involved and the potential for members on the scene being injured while performing actions required to mitigate the incident (fig. 3–13).[7]

Chief officers should not become involved in supervising routine tasks that can lead to tunnel vision, focusing only on the task at hand. They must be aware of all areas of the operation. Shifting from performing routine tasks performed in the past, to thinking as a manager responsible for everyone's actions, is not easy to do.

Use your firefighting experience and the four principles of risk management—identify, evaluate, prioritize, and control risk—to help you decide where the greatest risk to members might be and what actions you should be taking.

Fig. 3–12. Keep the incident commander informed of all risk management operations.

Fig. 3–13. Vigilant attention to changes on the fireground must be given.

Keep in mind the following:

- Risk a lot to save a lot. Keep significant risk to members limited to potential life-saving situations (fig. 3–14).
- Risk a little to save a little. The inherent risk of routine activities should be reduced and minimized.
- Risk nothing to save nothing. No risk to the safety of members is acceptable when there is no possibility of saving life or property (fig. 3–15).

Fig. 3–14. Risk a lot to save a lot (courtesy FDNY).

Fig. 3–15. Risk nothing to save nothing (courtesy FDNY).

Fire Chronicle

On October 11, 2006, the FDNY Memorial Day ceremonies had ended on West 100 Street in Riverside Park. While driving toward downtown Manhattan, we noticed heavy black smoke drifting across York Avenue. A report came in over the department radio for a fire involving a plane into a building on the east side of Manhattan. I turned to my aide and said, "Lights and sirens. We're responding in." The smoke was coming from East 72nd Street, a dead end block near the East River.

When we arrived, the only fire department unit on the scene was Engine 44, who were hooking up to a hydrant at the dead end of the block. They had transmitted a 10-77 on arrival (fire in a highrise residential building).

People were running in the streets away from the building, with the fear of 9/11 on everyone's faces

There were police officers with automatic rifles in the street. The initial reports had brought out a large number of emergency responders to the scene. I looked down the block and saw heavy fire coming out of two floors of the building (fig. 3–16). A section of the exterior wall was gone. It was obvious that some sort of collision or explosion had occurred. Debris was falling down and a small fire was burning near the lobby entrance. Looking at the debris, I couldn't tell if it was an airplane or helicopter.

I used the radio in Engine 44's cab to inform the dispatcher that the fire was at 274 East 72nd Street in a 40-story highrise apartment building, and that fire was coming out of four windows on the 29th floor and the floor above (fig. 3–17).

Knowing we would need more resources, I gave a 10-76 signal (fire in a commercial highrise building), which would bring additional units needed for this job (risk identification, evaluation, prioritization, and actions to control the risks).

Debris from the crashed plane was on fire and impeding access to the first floor lobby. I directed my aide to assist the chauffeur of Engine 44 in putting that fire out so incoming members could safely enter the building lobby (prioritized the risk).

Fig. 3–16. Plane crash into highrise building (courtesy FDNY)

Fig. 3–17. Fire damage on the 29th floor (courtesy FDNY)

In anticipation of the possibility of multiple casualties, I asked an NYPD officer to assist in keeping the street open for emergency vehicles. I also informed arriving EMS personnel to bring their medical equipment to the front of the building (risk evaluation, this was a low frequency/high severity incident).

I continued the risk assessment size-up. The front of the building revealed a 30-ft-long horizontal opening on the 29th floor, with heavy fire showing. On the 30th floor there was fire showing out five windows, with less damage to the exterior wall (fig. 3–18).

Fig. 3–18. Fire damage on the 30th floor (courtesy FDNY)

The area surrounding the damaged exterior wall appeared intact with the exception of the spandrel wall between the 29th and 30th floors, which showed heavy damage and large sections missing. There was no visible bowing or distortions in the exterior wall discernable from the ground level. Very early on into the operation I had received a report from a battalion chief (in a NYPD helicopter that was circling the building as per new procedures that were put into effect after 9/11) that the exterior wall appeared stable. With the exception of three windows on the 31st floor, which were cracked from the heat, all the other windows in the building were intact. There was no other visible damage to the building. I determined that the building appeared stable and did not have to be evacuated.

The possibility of a terrorist attack was part of my size-up. There were no reports of secondary devices or further explosions within the building or the immediate surrounding area. There were no indications of any effects from hazardous chemicals. No one was complaining of symptoms that might be associated with some form of toxic release. I was not receiving any reports that terrorism had caused the incident. Based on these conditions, I made a decision to use standard operating procedures for a fire in a highrise residential building.

The tactics would focus on protecting life in the immediate area of the fire and extinguishing the fire with an aggressive, interior, hand line attack.

At this time, NYC was still very much living with the possibility of another terrorist attack. I knew that some type of aircraft had crashed into the side of this building (it turned out that a single-engine aircraft carrying the Yankee pitcher Cory Lidle had crashed into the building on the 29th floor). But I didn't know what caused the crash. If it had been a terrorist attack and a secondary explosion or hazardous materials had been released, I would be second-guessing myself for the rest of my life, questioning if I did the right thing. Regardless of what could have occurred, the information I had during these initial periods of the incident and the risk management procedures put into action were the correct ones.

As the first alarm units began arriving, they were met with heavy fire on two floors with bricks and other debris falling down, smashing into the street near the building entrance. I set my command post up across the street in a small park, a safe distance from the debris collapse area. As members were given assignments, they were directed to keep looking up in order to watch the bricks that were falling. By keeping their heads up as they approached the entrance lobby, they would be able to doge the debris that was falling down in front of the entrance of the building (prioritizing the risk).

Battalion Chief Thomas Meara, from the 8th Battalion, arrived and I told him to establish the operations post in the fire area and operate as the attack chief.

As units reported in they were directed into the building and given assignments as per SOP for a fire in a highrise residential building. They heard the reports of a plane into a building and I could see the concern on their faces but never a hesitant step.

The first units in the interior reported heavy smoke in the 39th floor hallway. From the street I counted 29 stories to the flames out the windows. I asked my aide to verify the count, which he did. Not knowing the extent of the smoke conditions on the upper floors, I was concerned that the units were above the heavy fire that was showing on the 29th and 30th floors (fig. 3–18). I made radio contact with Engine 44 and Battalion 8, giving them this information. They acknowledged the message and shortly after transmitted that they had heavy fire on the 39th floor. Something was not adding up.

The fireproof construction and compartmented features of a highrise multiple dwelling would restrict heavy fire from spreading up 10 floors. It was obvious that some unusual floor numbering system existed. An accurate determination of the extent and location of the fire had to be made promptly.

The 9th Battalion was assigned as the building lobby command post chief. I requested them to verify the floor numbering system. An examination of the elevators revealed that the building had normal floor numbering up to the 10th floor, but then the 11th floor was designated as the 22nd floor. This 10-floor jump in the numbering of the building floors was announced over the handie-talkie radio, and units on the fire floor continued their attack on the fire. (It was later surmised that the higher the floor number, the more valuable the apartments, and this was a ploy to increase rents.) This numbering system was causing confusion as officers were informed to report into the operations post on the 38th floor. The only way to clear up this confusion was to inform all incoming officers of the existing floor numbering system (identifying the risk and taking actions to control it).

Chief of Department Salvatore Cassano arrived, I briefed him on the tactics being used, and he concurred with the strategy (fig. 3–19). He then took overall command of the fire and became involved in providing information and interacting with the army of multiple agencies that had arrived at the scene.

Fig. 3–19. Briefing the incoming incident commander on tactics and strategies (courtesy FDNY)

Approximately 25 min after units arrived on the scene, a hose stream appeared, shooting out of the 29th floor. The black smoke began turning white as the fire was being knocked down. A little later the fire shooting out of the 30th floor was turned into white smoke. FDNY Commissioner Nicholas Scoppetta was standing next to me in the street and exclaimed, "I can't believe the fire is out already."

It is hard for me to explain how proud I was of the members from the FDNY operating on that fire floor. No lives lost, property saved, and only minor injuries to operating personnel. It doesn't get any better than that for a fire of this magnitude, and that was all due to the aggressive attack and rapid control of the fire.

* * * *

We don't record the lives saved or the dollar amount of the property that we protect, but we should.

There were many severe risks to operating personnel at this event: building damage from the plane crashing into it, two floors of heavy fire weakening the structure, and the possibility of a terrorist attack adding unknown hazards. Knowing the construction type; identifying, evaluating, and prioritizing risk; and taking actions to mitigate the incident were all needed elements to effectively manage this situation.

Fireground Risk Management Objectives

Understanding a building's strengths and weaknesses, knowing how it dictates tactics, and using risk management to judge how to apply these tactics, provided a sound base for making judgments on the fireground.

To control risks, you must forecast the future of the emergency as it relates to member safety. Look for risks of immediate danger and evaluate the scene for risks that may present danger in the near future.

By looking ahead, imagining what might occur, it is possible to prevent accidents and injuries. Your intuition must be used to forecast the emergency situation and predict developments that can injure firefighters.

Not all injuries can be prevented during fire operations. Unknown dangers are always a possibility, and there will be times when situations will occur that cannot be controlled. But even during routine fires, when nothing serious goes wrong during an out-of-control fire in a structure, there were officers thinking ahead to avoid situations that could put operating personnel in peril. At a potential terrorist event with the possibility of exposure to chemical, biological, radiological, or nuclear exposure, the risk management plan must ensure that protective equipment and monitoring devices are available to members (fig. 3–20).

Fig. 3–20. Risk management must ensure proper protective equipment for possible chemical, biological, radiological, or nuclear exposure.

To ensure that a reliable risk management plan is maintained, a critique of incidents should be performed as a matter of routine. Information from these critiques should be used during periodic evaluations to determine what is working and what needs to be modified. Documentation of the updated, evaluated risk management plan provides reasonable evidence that risks have been identified and actions were taken to control risk in an appropriate manner. If an incident commander withdraws members from a building in danger of collapse and that building does not collapse, the incident commander was following good firefighting strategy by protecting life before property.

In many cases of studies of firefighter deaths, warning signs of building collapse were simply not acted on immediately; further confirmation or a second warning was anticipated. After the building collapses, hindsight benefits no one. Trust your instincts to determine the stability of the structure and use the principles of risk management to determine what actions are appropriate.

The tangible benefit of risk management on the fireground will be a reduction in the number and severity of accidents and injuries to members. Proper application of risk/benefit analysis will help ensure that the members who came to work go home from work.[8]

References

1. International Association of Fire Fighters, *IAF Fire Fighter Line-of-Duty Death and Injury Investigation Manual*, 2010.
2. FDNY Safety Bulletins 69 and 86, www.fdny.nyc.gov.
3. National Fire Protection Agency. *NFPA 1521: Standard for Fire Department Safety Officer*.
4. 2008 FEMA Incident Safety Officer, Instructors Guide. FEMA P-701/08162-2 Incident Safety Officer.
5. National Fire Protection Agency. *NFPA 1500: Standard on Fire Department Occupational Safety and Health Program* and *NPFA 1521: Standard for Fire Department Safety Officer*, 2008.
6. National Institute for Occupational Safety and Health (NIOSH). *NIOSH Alert: Request for Assistance in Preventing Occupational Fatalities in Confined Spaces*, NIOSH Publication No. 86-110, January 1986.
7. *NFPA 1521*.
8. *NFPA 1500* and *1521*.

Preventing Fire-Induced Building Collapses

4

Burnt-Out Buildings Lessons Learned

It bears repeating that no other professional organization witnesses more structural collapses than firefighters. Civil engineers, fire protection engineers, architects, and building code officials rely on accounts by eyewitnesses, video, and related studies to understand the dynamics of fire-induced building collapse. Firefighters are up close and personal when a ceiling collapses around their ears or a floor gives way beneath their feet. Highrise building fires highlight the weaknesses and strengths of modern construction and have been studied more than other structures because of the heavy loss of life and property damage when fires or earthquakes have caused collapse. Much of the information related to these building collapses document load shifts and progressive collapse.

The methods used in highrise construction and the fire protection systems in place will be analyzed for the following buildings:

1. The Beijing Mandarin Oriental Hotel fire in February 2009
2. The Madrid Windsor Tower Building fire in February 2005
3. The CESP Building on May 21, 1987, Sao Paulo, Brazil
4. The World Trade Center 7 Building fire on September 11, 2001

The Beijing Mandarin Oriental Hotel, aka CCTV Building

The multiple occupancy 34-story building in eastern Beijing was designed by the Dutch architect Sheeren Koolhaas and housed the Mandarin Oriental hotel, a theater, recording studios, and cinemas.

The building's decorative exterior of seemingly random diamond shapes was formed entirely by the primary load-bearing structural steel beams for the structure. In an interview with the building's designer, Sheeren Koolhaas, the process of creating the structure was described as follows:

> Making the scheme work required engineering of extraordinary intensity. Arup conducted separate computer models to predict how each of the 10,078 steel beams [exterior load-bearing members] could buckle or fail, and ran each of those failed-beam scenarios on four different pieces of software to crosscheck the results. To further vet the design, CCTV and the government convened a special oversight committee of top Chinese engineers . . . [and these were] the people who had written the codes we were breaking.[1]

A fireworks display caused the fire, which resulted in the death of one firefighter, seriously injured several others, and left the building a smoldering hulk. How the fire spread so rapidly is difficult to discern due to censorship by the Chinese government. It is still not clear how the fire ignited and spread into the structure on February 9, 2009. Fireworks could have melted the building's south side outermost wall, allowing sparks to pass through the metal wall and then set fire to the flammable insulation layer, from which the flames then spread inside the building. There were reports of a lot of debris inside that ignited very quickly.[2] Other reports say that the building burned so quickly and spectacularly because of the unique zinc/aluminum covering that gave it a shiny appearance.[3] Once the fire entered the building, an atrium opening in the area between the fifth and twenty-sixth floors produced a chimney effect, allowing the fire to spread rapidly through the interior.[4]

As of October 2011, the structure is still being rebuilt. On the plus side, as a result of this fire, additional funding was assigned to strengthen Beijing's firefighting capabilities.[5] If the total sum of these funds was put into fire prevention and firefighting capabilities prior to the fire, the entire city of

Beijing would be better prepared and the building would be a major tourist attraction, earning revenue for the city.

Fire engineers and architects make extensive use of computer modeling to assist in fire protection designs. I do not know if one of the 10,078 computer models performed included an analysis of a large amount of fireworks impacting on the building. But I doubt it. It would be impossible to predict every type of damage a building will suffer during the course of its existence. This unusual event highlights the limitations that must be recognized with regard to using computer modeling in defining fire protection. Extraordinary events occur in the real world and the proof is in the burnt-out shell of this highrise building.

Passive fire protection systems failed to inhibit the spread of this fire. I have no knowledge of the Chinese building codes or how the codes were applied to this building. I do have apprehension when code writers are subjected to political pressures to conform to the building owner's designs, which might have been the case with this building. The interview stated that the code writers became part of the architect's oversight committee: "the people who had written the codes we were breaking."[6] In this atmosphere, code writers can lose their sense of mission and might be swayed to decrease protection standards that can be expensive and/or hinder architectural designs.

The building was still under construction and did not have a fully functional sprinkler system. There were no effective extinguishment operations by the fire department. No active fire protection and the passive fire protection failure to limit the fire spread resulted in this building being completely burnt out and valueless.

The Madrid Windsor Tower Building

The 29-story Windsor Tower building in Madrid was completely burnt out as a result of a fire on February 12, 2005. Despite a total burnout and some local collapses, the strength provided by two "technical floors" (heavy concrete floors designed to give the building more strength) provided redundant structural support to distribute the loads caused by the fire damage. One was just above the ground level and the other at the 17th floor. This, plus the passive fire resistance of the building's concrete and steel with a central support system of concrete columns supporting concrete floors with steel perimeter columns, prevented the entire collapse of the building.[7]

The study of the fire concluded that the rapid fire spread to almost all floors was caused by two conditions: a lack of compartmentation, both vertical and horizontal, and no active sprinkler system. The collapse damage included a large portion of the floor slabs above the 17th floor, which progressively collapsed during the fire when the unprotected steel perimeter columns on the upper levels buckled and collapsed. The fire protection on the existing steelworks below the 17th floor had been completed at the time of fire except for the 9th and 15th floors. When the fire spread below the 17th floor, those protected perimeter columns survived, except for the unprotected columns at the 9th and 15th floors, which all buckled from the heat of the fire.

The reinforced concrete central core, columns, and transfer structures survived the severe fire conditions. The structural integrity and redundancy of these remaining parts of the building provided the overall stability of the building.[8]

This building resisted total collapse, for the most part, due to the concrete protected steel structures. However, the fire damage was so great that the entire building had to be demolished.

As occurred at the Beijing Mandarin Oriental Hotel, a lack of compartmentation, no sprinkler protection, and no effective extinguishment by the fire department resulted in a complete burnout of the building.

Sao Paulo Power Company (CESP) Building

On May 21, 1987, Sao Paulo experienced one of the biggest fires in Brazil, which caused a substantial partial collapse of the central core of the CESP Building 2. The National Institute of Standards and Technology (NIST) performed a study of this fire and reported the following:

This was a 21-story office building, headquarters of the Sao Paulo Power Company (CESP). Buildings 1 and 2 of this office complex were both constructed of reinforced concrete framing, with ribbed slab floors. Both buildings retained their original wood forms used for pouring the concrete floor slabs. Low-height plywood partition walls were also installed in the interiors.[9]

The ceiling in CESP 1 was made of plywood attached to the wood forms, while plaster tiles covered the ceiling in Building 2. Both buildings had automatic fire detection and manual fire alarms, but *no automatic sprinkler system*. Six footbridges connected these two buildings to permit convenient pedestrian access at various levels.

The CESP fire started on the 5th floor of Building 1 from electrical causes, and progressed rapidly upward due to the lack of compartmentation and the combustible (wood) ceilings. The severity of the Building 1 fire caused it to extend into Building 2.

CESP Building 2 had similar combustible partition walls and wood floor formwork that ignited simultaneously on several floors due to auto exposure from the high thermal radiation from Building 1.[10]

Firefighting efforts in both buildings were unsuccessful. Approximately two hours after the beginning of the fire in CESP Building 2, its structural core area throughout the full building height collapsed. This collapse was attributed to the thermal expansion of the horizontal concrete "T-beams" due to fire exposure, which led to the fracture of the connection points of the vertical framing elements in the middle of the building. This resulted in the subsequent progressive loss of their load-carrying capacity.

The load was shifting beyond nominal load-bearing limits and, like a ship taking on water as it begins to sink, further decreasing its buoyancy, the highrise building's main structural load-bearing elements took on more and more loads and began to progressively collapse.

When Building 2 collapsed it increased the fire conditions in Building 1, the original fire building, which was burning out, by reigniting floors 1 through 4 due to the flaming debris falling on the lower levels.

It was reported that this entire incident, from the original fire ignition in Building 1, spread to Building 2, the collapse of Building 2 core, and the final reignition and burning of the lower floors of Building 1, lasted a total of about seven hours.[11]

The NIST study presented these conclusions:

> Given that there can be no guarantee that a fire will not occur in a given building, or that it will be successfully contained and suppressed, the fire resistance of the building structure must be duly assessed in its design in order to avoid local and progressive collapses. Since several of these documented cases demonstrated various member and structural connection failures, a better understanding of the response of various building connections to fire is needed.[12]

This investigation indicated that connection point failure often initiates progressive collapse.

World Trade Center 7 Building

The WTC 7 building was reported as the first time a fire caused the total collapse of a steel-framed highrise (greater than 75 ft) building. The full collapse of the 47-story steel-framed WTC 7 occurred approximately eight hours after the collapse of WTC 1. The structure, built in 1987, consisted of perimeter steel frames, two-story belt trusses, and an interior braced core at the lower levels provided lateral resistance features. The floors were typical steel beams with composite deck and concrete topping.

The collapse of the nearby WTC towers broke the city water main, leaving the sprinkler system in the bottom half of the building without water.[13] The upper floors had a captured sprinkler reserve water supply that supplied water to the sprinkler system on those floors during the fire.

The report indicated that a critical steel support column failure on the 13th floor led to a progressive collapse, resulting in complete catastrophic failure of the building.[14] Column failure bringing down the entire building is a disaster that firefighters understand all too well.

The NIST investigation presented these findings:

- There was no redundancy in the source of water supply for the sprinkler system in the lower 20 floors of WTC 7. Since there was no gravity-fed overhead tank supplying these floors, the sprinkler system could not function when the only source of water, which was from the street mains, was not available.

- In current practice, architects typically rely on cataloged *ASTM E119* (American Society for Testing and Materials) test data to specify the required passive fire protection that is needed for the structure to comply with the building code. They are not required to explicitly evaluate the fire performance of the structure as a system (such as analyzing the effect of the thermal expansion or sagging of floor beams on girders, connections, and/or columns). Structural engineers are not required to consider fire damage as a condition in structural design.
- Fire protection engineers may or may not be called upon to assist the architect in specifying the required passive fire protection. Thus, none of these professionals has been assigned the responsibility to oversee the structure's entire fire protection systems to ensure the adequate fire performance of the structural system.
- There is a critical gap in knowledge about how structures perform in real fires, particularly considering the effects of the fire on the entire structural system, and the interactions between the subsystems, elements, and connections.[15]

The collapse of WTC 7 was *a progressive collapse, a common factor in most fire-induced building collapses*. The number one recommendation from the report recognizes the importance of addressing this risk:

> Progressive collapse can be prevented in buildings through the development and nationwide adoption of consensus standards and code provisions, along with the tools and guidelines needed for their use in practice. . . . [A] standard methodology [should] be developed—supported by analytical design tools and practical design guidance—to reliably predict the potential for complex failures in structural systems subjected to multiple hazards.[16]

Multiple hazards include fire damage load stresses as well as wind, snow, and other design-considered nominal loads. If these recommendations were acted on, it would greatly add to our understanding of how a building might collapse as a result of fire damage. Further recommendations called for improved compartmentation to limit the spread of fires.

The fire-induced collapse of WTC 7 was analyzed by many agencies. Findings from the analysis performed by the American Society of Civil Engineers also clearly points to what needs to be done:[17]

- Buildings should be designed with sturdy backup structural supports to bear the weight held by the primary supports when damage to the building occurs.
- Fireproofing needs to adhere under impact and fire-induced steel deformation so that the protective coatings remain on the steel and provide the intended protection.
- When sprinkler systems are a critical part of a building's fire protection system, the water supply should be reliable and redundant.

The two primary considerations in fire-resistant construction are design and materials. A building should be subdivided by fire-resisting walls, floors, and partitions to limit the spread of fire. Elevator and stair shafts, atriums, and other vertical structures within the building must be isolated by heavy fire-resistant walls because vertical openings act as chimneys, increasing the intensity of a fire. All doorways or other wall openings in exit corridors should be provided with doors that are self-closing or that close automatically in the event of fire. Materials used for interior finish, particularly in exit corridors, should be able to resist the spread of fire, and the codes detail minimum flame-resistance requirements. There are three ways to protect load-bearing structural members from the effects of heat:

- Encasement
- Spray-on fire protection material
- A floor with either of these protections with a membrane ceiling underneath it

Steel encased in cement will generally allow the load-bearing capabilities to withstand the thermal effects of heat generated in a typical building fire and has a reputation as the most effective method to protect load-bearing structural members, period. All four of these building fires had two common themes:

1. There was no water on the incipient fire due to a lack of active firefighting efforts and/or no sprinkler system activation.

2. A lack of effective passive fire-resistant barriers and design features allowed the fire to spread through the building, resulting a complete burnout and significant or total structural collapse.

Testing of Fire-Resistant Construction Materials

The lessons learned from fire-induced collapse and burnt-out shells of buildings is that adequate compartmentation and effective fire protection of structural members should be mandated. The safety features in our buildings and homes must be able to withstand the temperatures of real-world fires long enough to permit safe egress and reasonably allow the structure to survive the effects from the fire.

Building codes can require where fire-resistant barriers are mandated, but how do we know if the structural materials used can provide adequate fire protection? The testing methods applied today often do not reflect the dynamics of the synergistic effects fire has on structural materials and often do not test the materials up to the failure point. This reduces the testing value as it pertains to a structural elements fire-resistive rating. The testing of fire-resistant materials typically uses a similar but single fire exposure (e.g., *ASTM E199* and *ISO 834*) with only a limited amount of structural elements subjected to more than one side testing, such as columns and beams.

I have had many conversations in firehouse kitchens with fellow firefighters, some of whom were also civil engineers, about this subject. It was evident to us that having one side of a fire-resistant structural material exposed to a controlled heating process, kept at an exact temperature for a specific time, was not what occurred during actual fire conditions (fig. 4–1). It is hard to believe that this is accepted as the norm to reflect credible testing results. Yet that is exactly what is occurring.

It is a matter of common sense to firefighters that several sides or even the entire area of a structural member's connection point assembly can be heated at varying temperatures during a fire.

Fig. 4–1. Fire will affect several or all sides of a structure (courtesy FDNY).

The National Institute of Standards and Technology (NIST) recognized the need to analyze fire resistance testing. Their report explicitly described the need for a revision in the way this testing is performed:

> It has been assumed in standard tests, that the actual fire will expose one side of an assembly. A building fire will often impinge on several if not all surfaces of a structural member. The existing fire test facilities evaluate building elements with specific size limitations that cannot accommodate significantly larger building elements. The performance under a specific load on a shorter member will not reflect the same loading on a longer member. Other loads or stresses may also be imposed on a structural element that may affect its structural performance during a fire event are not tested. Loading with respect to tension and shear are not typically done.[18]

The report goes on to highlight the limits of present day testing of fire-resistant materials:

> The standard practice in fire resistance testing is to test each building element individually. For example, roofs or floors are not tested in combination with walls or columns. Thus, the roof assembly is tested as an individual element and the column or the wall is also tested as an individual item. Connections between building elements are not necessarily evaluated under the existing fire resistance test methods. In some cases, such as with floors, the ends of beams may be restrained, but they are typically not exposed to the actual fire environment.[19]

The load testing of building structural members such as beams, girders, and columns are not accurate because the process used in the test does not take into account the additional nondesigned loading that fire-weakened members can exert on other structural elements. When the combination of thermal load effects and structural load-carrying performance is ignored and/or underestimated, how can code officials accurately judge what is required to protect building occupants?

One of the most common structural elements that initiate a fire-induced progressive collapse is the connection point. To effectively judge the stresses applied to the connection points during load testing, the stresses on the fire-weakened interconnected structural members should be measured. This type of data will provide a level of understanding that is required for the development of systems that would resist movement that might initiate a progressive collapse event. It could also indicate how much distortion a structural element can withstand before being destabilized to the point of failure. This information would be of significant value to firefighters attempting to visually judge the potential for failure of a connection point. Communication between architects, structural engineers, firefighters, code officials, and federal agencies like NIST is a necessity if we are to determine what type of testing accurately reflects real-world fire situations. A greater understanding of the impact that fire has on the building elements must be achieved.

NIST should be recognized for their efforts in performing a scientific enquiry into the faulty testing systems and exposing the inaccurate methods that have resulted in a lack of accurate information regarding fire-resistant materials. They have performed a great service to all firefighters.

Defining Fire Resistance by Using Technology

Fire protection engineering is the application of science and engineering principles to protect people and their environments from the destructive effects of fire and smoke. Fire protection engineers identify risks and design safeguards that aid in preventing, controlling, and mitigating the destructive effects of fires. Fire protection engineers assist architects, building owners, and developers in evaluating the buildings' life safety and property protection goals.

As chief of fire prevention, I was often asked to review new building construction and approve fire-engineered solutions that modified existing New York City fire and building codes. Most of these solutions involved the ability of having fans, automatically activated by smoke detectors, remove smoke from a sprinkler protected area. This system design replaces passive fire protection and compartmentation.

Atriums are an example how "smoke control" might be applied in lieu of fireproof compartmentation. Atriums eliminate the passive fire resistance of a floor assembly to contain fire and smoke. In order to create a large, open interior atrium space inside a building beyond the compartmentation limits enforced by the building code, a smoke control system can be installed to draw the smoke out, removing the contaminants from the open atrium area. Passive fire-resistant walls and enclosures are replaced with *air* movement. Often, in highrise hotels, occupant exit passageways are in the open atrium area.

Fire Chronicle

I was captain in a ladder company working at a fire that involved the contents of a 20 × 25 ft general storage room in a highrise hotel building, with a multistory open atrium. In this building the "smoke control" systems had been activated by smoke detectors in the storage room. Smoke ejector fans were removing the products of combustion up toward the roof level. There was a residual, light smoke condition on the lower floors. A partial evacuation of the building was in progress. Occupants were using exit passageways in the open atrium area. While the smoke condition was not a problem for the occupants that I contacted on lower floor passageways, it was noticeable that the system did not completely remove the products of combustion. This indicated that there is a period of time during fire conditions and/or immediately after extinguishment where the products of combustion will be present in occupied areas during smoke removal that a fully functioning system will not eliminate. I did not stay in the hotel atrium to witness how long it took for total smoke removal from all levels of the atrium.

This fire resulted in occupants using hallways in open atrium areas exposed to the products of combustion during active fire conditions. This is a consideration that should be acknowledged when replacing floors and walls with *air* movement.

On a side note, an announcement was being made by the fire safety director that the fire department was on the scene and in the process of controlling the fire. This was done in a calm, professional manner that impressed me as a good example of how to keep panic under control. That was until I came across a crowd of confused, worried hotel guests from Japan in a hallway who couldn't understand what was being said but knew a fire was in the building. It is these types of unpredictable events that often occur during real-world building fires.

* * * *

I like atriums as a design aesthetic. To walk into an expansive open area inside the lobby of a highrise building is exhilarating. But they do present a serious threat of smoke and fire permeating the interior of the building. If you are going to build atriums, there should be substantial passive fire-resistant materials enclosing the atrium space. Also, required egress pathways should be enclosed.

Full-scale tests involving large open area spaces with instruments recording heat and real fire-generated smoke levels, throughout the structure, are required to accurately record the data needed to design an effective atrium smoke control system. As I witnessed in this newly constructed highrise hotel, initially, as the fire was being extinguished, some smoke remained inside the building while the smoke control system was operating.

Compartmentation and sprinklers

The major issue with relying on sprinkler systems to replace compartmentation is, what do you do if the sprinkler system is out of service? To my knowledge no fire protection engineer has proposed a plan requiring evacuating various portions of the building anytime the sprinkler system, or a section of the system, is shut down for repairs or other causes. When fire-resistant compartmentation has been replaced by smoke control systems and sprinklers, planned building evacuation should be a part of the design.

Fire Chronicle

On June 12, 1979, a fire during business hours was reported in Macy's department store on 34th Street in Manhattan. Arriving members from Ladder Company 24 found a light smoke condition on the sixth floor. The fire was hidden in the storage room and the smoke in the ceiling plenum area was not visible to shoppers. Conditions changed suddenly when the hidden fire flashed over the uncompartmented sixth-floor ceiling with pressurized black smoke rolling across the area that quickly banked down to the floor level, completely obscuring visibility. The expanding heavy smoke caused windows to explode and created a heavy fire condition on the sixth floor. Firefighter Walter Smith of Ladder 24 was unable to get out of the fire area and perished. The sprinkler system on the floor was out of service for repairs and did not operate on the fire. The fire required a fourth alarm response to complete extinguishment.[20]

* * * *

Macy's was open to shoppers when this fire occurred. As a result of this tragic fire, the FDNY set up procedures for when a sprinkler system went out of service in a building. The fire department must be notified and the local fire company is required to promptly inspect the premises and issue a violation order to restore the system. The paperwork generates a reinspection by the same unit, ensuring that the system has been put back in service as required.

This fire prevention system for out of service sprinklers can be found in the *FDNY Fire Prevention Manual*, form A-500.[21]

It should be noted that in NYC codes, for certain occupancies with a sprinkler system out of service, the administrative battalion chief has the authority to order a vacate of the affected area if deemed necessary. Just because a sprinkler head activates, it does not mean that the fire will be extinguished. It is not uncommon to have the area under the activated sprinkler head obstruct the water disbursement, preventing total extinguishment.

Fire Chronicle

Engine Company 26, located on West 37th Street off of Seventh Avenue in Manhattan, in the heart of the "Garment District" was the first firehouse I was assigned to as a probationary firefighter. Sweatshop factories where raw textiles were cut into garments were common in this area. Many of my initial fires involved these types of occupancies. Often careless smoking would ignite the scraps of cloth that littered the floors. The ceiling mounted sprinkler heads would activate but could not extinguish the fires burning under the cutting tables that ran the length of the factory floor.

Typically we would stretch a 2½-in. hose line off the standpipe outlet into the factory floor where very heavy smoke obscured a large open area. We would crawl down the narrow aisle space between the cutting tables, with highly flammable, plastic-covered garments hanging on poles overhead, until we located the fires burning under the tables that the overhead sprinklers could not extinguish. It would often take two or three trips down the entire length of the floor and back until we located the burning fire.

* * * *

I am a firm believer in sprinkler systems. They are a major factor in saving lives and protecting property every day and night. I strongly advocated requiring them in all types of structures and occupancies. But they often do not extinguish a fire. The problem is when they are used to reduce other fire protection measures and reduce the overall effectiveness of the building's fire protection.

When a building has replaced passive fire-resistant barriers with sprinklers, and the sprinkler system does not totally extinguish the fire, the potential for significant fire damage and loss of life from the effects of toxic smoke can be increased to an unacceptable level. Appropriate fire protection requires a balance, combing sprinkler systems with a level of fire and smoke barriers that can prevent a catastrophe. The only other alternative is to have a preplanned evacuation of the building or areas of the building whenever the sprinkler system is out of service.

Another building safety feature that impacts on the potential for fire-induced collapse is the egress system. What does an interior stair design have to do with building collapse?

In highrise buildings, a limited evacuation of building occupants is necessary to allow firefighters to perform extinguishing efforts without being overcome with required rescue attempts or having to delay extinguishing efforts because occupants are attempting to use a fire attack stairway to exit the building. If firefighting personnel cannot efficiently achieve an aggressive interior attack on the fire, the fire will expand to the point where structural damage occurs.

Fire towers

Fire tower stairs have a long history of providing a safe, smoke-free environment for egress during a fire. This stairway design has a fire-rated enclosure with a vented vestibule that has self-closing doors. The stairway also has self-closing doors. The combination of two separate doors, walls, and an enclosed vented vestibule effectively block smoke and heat from entering into the fire tower stair, leaving it clear for occupant egress in an emergency.

I have operated at many building fires under heavy smoke conditions where these stairs were relatively free from any products of combustion. I am absolutely certain that they provide the safest emergency evacuation stairs for building occupants.

Fire tower stairways, or smoke-free stairs as they are also known, are specifically designed *for building occupants' emergency egress* in highrise buildings. Fire departments cannot use a fire tower for an attack stair due to the movement of air toward the open stairway and open vestibule doors when advancing a hose line into the fire area. If both of the stairway door and the occupancy door are open at the same time, the natural air movement draws smoke and heat toward the stair tower and the hose line nozzle team at this doorway. For this reason, during highrise firefighting operations, these stairs are prohibited from being used as an attack stairway. The attack stair must be a standard exit stair other than a fire tower stairway.

This works out very well when fire conditions require partial evacuation of the building. Occupants use the fire tower evacuation stair for safe egress and the fire department uses a different stairway for fire operations.

Fire tower stairways are being replaced with "pressurized stairways." Pressurized stairways use fans to create air pressure in the stairway that resists the movement of smoke from the occupant side of the doorway into the stairway, thus leaving the stairway clear of smoke.

There have been many documented tests that validate the effectiveness of this system.[22] The problem with the test that I have reviewed is that they test using "non-fire/fire conditions," which means that a real fire was not used to create the smoke conditions for this test. Why not? If you are going to replace a fire tower, a proven safe evacuation stair, with a less expensive engineered solution, the test used must reflect accurate fire conditions as much as possible to have valid information. Once again, testing of fire protection is less than what should be done.

My experience with pressurized stairways indicates that when too many doors are opened by occupants, and/or firefighters are performing searches for victims, a pressure drop occurs within the stairway, resulting in a lack of an effective smoke barrier.

Most of the tests that I reviewed limited the number of door openings during the testing procedures. My question is what happens to the air pressure smoke barrier when more than the set testing limits of stairway doors are open? There is no definition of how many open doors are required to be used in tests to prove that the system will provide an effective smoke barrier. It is impossible to prevent multiple stairway doors from being opened by occupants fleeing from a building fire. This problem has been noted in the investigation of the Bankers Trust Building fire. "It is very difficult to design a system that can balance the pressurization against multiple doors being open at the same time, and also account for stack effect and pressures resulting from the expanding air mass created by a hot fire."[23]

I have operated at many highrise fires where announcements were made over the building's speaker system telling occupants what stairway to use, but this did not prevent occupants from opening other stairway doors in an attempt to evacuate the building. In reality, building occupants opening a variable number of stairway doors is an inherent weakness in this system, which cannot be tested for in any reliable manner.

Overpressurized stairways also present problems. I have experienced difficulties while performing searches on upper floors where the stair pressurization is so great that it makes opening stairway doors from inside the stairs very difficult. Also, there have been times when the noise of the fans in the stairs is so loud that it inhibits communications. Fire towers have none of these drawbacks.

The reason fire tower stairs are very rarely installed in new construction is that these types of stairs take up more space than normal stairs and are more expensive to build. The loss of rental space and additional cost make them "unattractive" to building designers.

Using computer models that do not apply accurate variables in determining the effectiveness of pressurized stairways demonstrate the possible faults and drawbacks associated with many fire-engineered solutions. Relying on systems that have faulty testing methods is a major concern when using *air* to replace structural fire-resistant barriers. Most fire protection systems using smoke detectors, sprinklers, and smoke ejector fans are satisfactory and can provide an acceptable level of safety to building occupants, *when the fire is contained by the sprinklers as designed*. When the fire spreads beyond the system's containment capability, due to a malfunctioning sprinkler system or other causes, the open space area and the air movement caused by the smoke removal fans might make matters worse by creating conditions that could intensify a fire.

These systems are complex. There are many events that fire alarm systems activate, including closing fail-proof smoke doors, stair pressurization, elevator recall, alarm signals being transmitted, closing air ducts, and so forth. A weak link in a complex system is more prone to failure when taxed by damage from fire. Any type of system malfunction can have catastrophic consequences; if the smoke alarm system does not trigger activation of the smoke ejector fans, toxic smoke can rapidly spread inside the building and quickly escalate to life-threatening levels.

I have not witnessed a heavy smoke condition generated by a large fire area in an atrium area. I am sure there have been hundreds if not thousands of computer models that have proven the systems involved would perform as advertised. The concern is that no accurate large-scale burn test has been performed to verify these computer model results. Presentations involving computer models are very sophisticated and impressive. It is difficult to be critical of the logic presented at meetings when complex systems like smoke control solutions are proposed. As chief of fire prevention, I had a policy where no decisions would be made at an initial meeting; it was strictly to hear the proposals.

We would then have the time to examine the computer models. I was fortunate to have a staff of extremely bright, persistent technical examiners who could understand and evaluate the information contained in these computer models and glossy graphic publications.

Often there were gross underestimates of the effects that a fire might have. For example, a fire model exhibited a fire in a location where, due to design features, it would have the least effect on smoke travel or fire spread in the building. Moving the fire scenario to another location caused the data on smoke movement to show a more significant negative impact. Also the level of fire was often based on minimum events. When these minimum levels were exceeded, often the smoke and heat levels generated were unacceptable. This type of detailed analysis of proposed building designs is not available to most departments, and there was a limit on how many projects I was able to have investigated and reported on. When I was first introduced to these "smoke control solutions," I realized that these methods were moving in directions that could present problems for fire service personnel. Many of the chief officers I have discussed this with are very concerned about having air movement by fans, automatically activated by smoke detectors, introduced into the area of an out-of-control fire inside a building.

As buildings become more airtight structures to accommodate HVAC systems, the air movement caused by HVAC systems, fire conditions, stack effect, and by smoke purge fans when activated by an alarm system creates a very complex arrangement that challenges the fire chief's understanding of how the building systems are affecting the heat and smoke conditions generated by the fire.

There is a certain "living" quality about an *uncontrolled fire* that creates a uniqueness where one fire will not develop and grow in the exact same way as another one (fig. 4–2). Similarities exist regarding the volume of smoke and flame activity of particular materials, but frequently results occur that were not anticipated or designed for. The synergistic effect of the various combustible products and the amount of available air involved in uncontrolled fires results in produced gases with thermal capabilities that are very difficult to analyze with a computer model.

Fig. 4–2. Uncontrolled fires are unique in that no two fires will develop and grow in the same way.

Fire Chronicle

There are times when a firefighter will pass limited fire areas to make a search for life. I have entered rooms where a couch is actively burning but areas beyond it are not involved in fire. I have witnessed fires where one burning couch could be passed to perform a search in relativity safety while another flared up, igniting the entire room in seconds (and had me diving for the exit door).

* * * *

Improbable events with unforeseen effects can and do occur during uncontrolled fire situations. This limits the value of computer modeling.

The simple truth is that we do not have a complete, accurate understanding of the dynamics of an out-of-control fire inside a building. To ignore this is to put building occupants at peril. Talking with fire protection engineers, I proposed that the fans be activated only when a sprinkler head is discharging water in order to ensure that water is being applied to the fire area before the air movement is introduced. This would provide some assurance that the fire would not be intensified by the "fire protection" systems. The engineers I spoke with noted the concerns and said they would look into what type of device should activate the fans. I am not aware of any actions that were taken to control this specific type of risk.

Like the code of ethics for physicians, "First, do no harm," a no harm approach should also be applied to fire protection systems. You do not want to make matters worse by having fans activated that may draw fire or smoke into clean, uncontaminated areas or have systems fail to protect building occupants due to a lack of compartmentation.

Code writers continue to rely on the assurances that a computer model system can mimic an actual fire event. In fact, they can be getting a very limited view of the dangers that a real-world fire presents. Today there are many engineered structural products with less mass. If you look at new one- and two-family homes under construction you can see right through them. The mass of structural members that was once present in older type construction has now been replaced with *air*. Geometric trusses and lightweight metal frames are the new replacements for the redundant safety of structural mass fire resistance in smaller buildings. This results in providing less resistance to fire-induced building collapse. The solution is to use fire protection methods such as sprinkler systems, smoke detectors, and smoke control systems, combined *with* passive fire-resistant barriers that have proven to be effective under actual fire conditions.

With this solution there must be a balance struck between cost of safety design features and construction materials.

Highrise office and residential buildings have an excellent safety record. For the most part this is due to the installation of sprinkler systems. Sprinklers have proven to be reliable, but failures do occur. Tracking sprinklers systems that are out of service (OOS) by using the FDNY procedures that were outlined earlier can help to limit the amount of time the building and occupants are unprotected and can reduce this hazard.[24]

This being the case, how do you justify requiring an increase in passive, fire-resistant compartmentation and enclosed egress passageways that could cost millions of dollars? How can a calculation be made on the amount of lives that might be saved each year?

One factor to consider is that buildings are being built with less structural material than before and are built faster, which results in a significant reduction of personnel hours and construction material costs (fig. 4–3). I am not aware of any research that has compared the cost factors of newer, lighter weight building material and reduced labor hours during the construction phase to the cost of buildings built before the rapid development of engineered construction products. I believe the difference in these relative savings during the construction phase is considerable. If this is accurate, it translates into more profits being made by building owners.

Some of these additional profits could be directed toward paying for increasing passive fire-resistant construction features, providing a balance directly related to safety and construction methods. It is a direct link to the construction industry and investor profits and should be recognized in some manner.

Fig. 4–3. Buildings are being built faster with less structural material.

Studies of Fire-Induced Building Collapse

Before examining the current thinking on how to prevent fire-induced structural damage and building collapse, a review of the history of fire protection will provide a framework for the major concerns.

One of the first persons to address the need for fire protection in highrise buildings was FDNY Chief of Department John Kenlon. With an explosion of new construction in New York City in the early 1900s there were many opportunities to observe situations where things went very wrong. When the Equitable building burned to the sidewalk in 1912, it became apparent to Chief Kenlon that a new technique of fire-resistant construction was required. His solution was to recommend building codes that would require a building's structural elements to withstand a temperature of 2,000°F for four hours without suffering a significant collapse of any kind.[25]

The rebuilt Equitable building was built close to those standards. Not long after it was completed it suffered a gas-fueled fire that extended in a shaft from the basement to the 35th floor, where it mushroomed out. At the time it was the highest fire in a building. The fire was extinguished from the standpipe system that had two reserve water tanks. As highrise buildings began sprouting up during the 1920s, fires on upper floors increased and with them came new logistic problems for firefighters. If fires above the 15th floor were to be fought in a rapid manner, elevators would be needed to provide firefighters with the capability of getting equipment and personnel to the fire floor.

Chief Kenlon went on to express that the single element that was dreaded the most was the collapse of a supporting column. A recent complete collapse of three cast iron columns which brought down an eight-story building was cited as the reason for the need of fire-resistant material. Chief Kenlon stated, "Had these columns been of steel covered by two inches of cement plaster or other fire insulation, firmly anchored to resist fire and water, the fire would have been just a simple packing room blaze, extinguished without difficulty."[26]

The need for compartmentation by using fire-resistant structural materials in highrise buildings was evident in the 1920s. A three-story warehouse fire on West 59th Street had two huge unbroken floor surfaces, one 50,000 sq ft and the other 19,000 sq ft, divided only by a single brick wall. There were new automobiles stored on the floors. Gasoline ignited, causing the fire to spread rapidly by auto exposure from the windows on the fire floors to upper floors before adequate extinguishing efforts could be put in place. The building and all its contents was completely destroyed by the fire. Kenlon believed wired window glass was one product that could help prevent the flames from spreading from floor to floor.

Chief Kenlon was a realist about fire prevention. "I believe human nature and modern society are so constituted as to make the literal prevention of fires impossible. Granted then, that fires will always get started somehow, it becomes our task to prevent them from spreading." He had high regard for the new sprinkler systems just being installed saying, "I can truthfully say that where there are no sprinklers, there is not adequate fire protection."[27]

Chief Kenlon knew that fires will occur in spite of public fire prevention campaigns or other similar preventive measures. He pointed out the ways to protect buildings from fire-induced collapse: compartmentation using passive fire-resistant materials combined with active fire-resistance systems (sprinklers) and the need to protect load-bearing structural elements with effective fire-resistant materials.

These same conclusions are being substantiated in studies being conducted today with a focus on how to prevent a "progressive collapse" of a building.

There are ways to design buildings to create resistance to the destructive effects of fire and reduce a structure's vulnerability to progressive collapse by incorporating redundancy and local resistance to nondesigned stresses.

These methods have been supported in a number of different studies on the subject. Some of the studies focus on the damage caused to a building's structure by earthquakes. This information can also be applied to the similar type of destruction of structural elements caused by fire.

In a study performed by the American Society of Civil Engineers, redundancy is noted as a significant element in reducing structural failure:

> Buildings and other structures shall be designed to sustain local damage with the structural system as a whole remaining stable and not being damaged to an extent disproportionate to the original local damage. This shall be achieved through an arrangement of the structural elements that provides stability to the entire structural system by transferring

> loads from any locally damaged region to adjacent regions capable of resisting those loads without collapse. This shall be accomplished by providing sufficient continuity, redundancy, or energy-dissipating capacity (ductility), or a combination thereof, in the members of the structure.[28]

Redundancy in concrete columns by closer reinforcement spacing can provide resistance to structural failure due to an earthquake and would provide the same type of support during fire-induced collapse incidents.

The problem is that the degree of redundancy is not specified. The lack of understanding on how to define structural redundancy undermines the protection offered by a code or standard. Architects and engineers are left to their own interpretation as to what constitutes a sufficient amount of redundancy.

The American Concrete Institute has provided extensive requirements for structural integrity and their code commentary section makes reference to improving redundancy but again fails to provide explicit information on how to apply redundancy or alternate load paths in the Code.[29] This deficiency in providing collapse protection was noted in a report on progressive collapse basics by Shankar Nair's paper titled "Preventing Disproportionate Collapse," section V.a., Redundancy or Alternate Load Paths:

> In this approach, the structure is designed such that if any one component fails, alternate paths are available for the load in that component, preventing a general collapse from occurring. This approach has the benefit of simplicity and directness. In its most common application, design for redundancy requires that a building structure be able to tolerate loss of any one column without collapse. This is an objective, easily understood performance requirement. The problem with the redundancy approach, as typically practiced, is that it does not account for differences in vulnerability. Clearly, one-column redundancy when each column is a W8 × 35 does not provide the same level of safety as when each column is a 2,000 lb/ft built-up section. An explosion that could take out the 2,000 lb/ft column would likely destroy several of the W8 columns, making one-column redundancy inadequate to prevent collapse in that case. And yet, codes and standards

> that mandate redundancy do not distinguish between the two situations; they treat every column as equally likely to be destroyed. In fact, since it is generally much easier to design for redundancy of a small and lightly loaded column, redundancy requirements might have the unfortunate consequence of encouraging designs with many small (and vulnerable) columns rather than fewer larger columns. For safety against deliberate attacks (as opposed to random accidents), this may be a step in the wrong direction.[30]

The knowledge to prevent fire-induced building collapse is here. The need is how to define what constitutes "redundancy" in a code. Fire departments across the country should bring this topic to the attention of the construction industry in an effort to initiate effective codes that would cause the reduction of fire-induced building collapse through column redundancy. I know Chief Kenlon would approve. In fact, as NIST has reported, there have been codes written to address this issue. In 2000, the GSA *Facility Standard for Public Buildings* created a prescription code to prevent progressive collapse regardless of the type of damage inflicted on the structural members, and it provided a substantial safety level.

> The structure must be able to sustain local damage without destabilizing the whole structure. The failure of a beam, slab, or column shall not result in failure of the structural system below, above, or in adjacent bays. In the case of column failure, damage in the beams and girders above the column shall be limited to large deflections. Collapse of floors or roofs must not be permitted.[31]

This type of specification eliminates any conflicting interpretations of the code intent and ensures adequate requirements to limit structural failure.

Unfortunately, this code was changed when the 2003 edition of the GSA's *Facilities Standards for the Public Buildings Service* came out. It retained the "Progressive Collapse" heading from the 2000 edition, but replaced all of the section related above with the statement: "Refer to Chapter 8: Security Design." Chapter 8 provisions only apply to buildings deemed at risk of blast attack. For these buildings, the chapter provides general performance guidelines and references to technical manuals for study of blast effects.[32]

This change deleted a code that reflected the concerns of the fire community and provided clear guidelines on requirements to improve building stability.

Connection points

Effective interconnections creating load continuity can improve a building's resistance to collapse. Recent building collapses show that failure could have been avoided or at least reduced in scale if structural components had been interconnected more effectively. This is the basis of the "structural integrity" requirements in the ACI 318 specifications.[33]

The paper titled "Progressive Collapse Basics" by Shankar Nair, written in 2004, pointed out the concerns about cataclysmic collapse similar to those Chief Kenlon had with collapsing columns over 80 years before. Here we are in the beginning of the 21st century still waiting for the codes to catch up to the reality of prescribing fire protection that can prevent building collapse.

Increasing Firefighting Capabilities

What can the fire service do to help prevent fire-induced building collapse? Ensure adequate *water supply*.

Obviously the major cause of fire-induced building collapse in buildings is when the sprinklers and/or firefighting efforts are unsuccessful in extinguishing the fire. This is true regardless of the building's height. Fire departments must have the ability to fight a fire in any area and on all levels of a structure. If adequate water is not available at sufficient quantities, firefighters are regulated to being observers to the tragedy.

When it comes to influencing lawmakers on the need to provide fire protection, fire departments must play a large role in defining what needs to be in place in order to extinguish an out-of-control fire inside an occupied building. This is an essential fire department requirement that must be met.

My responsibility as chief of fire prevention was to advise the department on all matters concerning building and fire codes during a code revision process. We viewed this mandated code review as an opportunity to improve the standpipe system for highrise buildings in NYC. I gathered members of the FDNY together, Battalion Chief Thomas Meara in particular, to help design a reliable standpipe system that could provide an adequate supply of water on upper floors of a building. Some of the requirements we decided on went beyond what most other codes called for.

The system we wanted would provide water supply with sufficient redundancy to a standpipe piping system that could deliver the water at the required quantity and pressures anywhere in the building.

In particular, we wanted to ensure that the water supply in super highrise buildings, those over 600 ft high, could support the extinguishing efforts of our firefighters.

One of the major problems with water supply in highrise buildings is the failure of the standpipe system's piping when supplying a building standpipe with water from fire department apparatus pumpers for fires on the upper levels of a building. A frequent location of piping failure is on the building side of the standpipe siamese connection, where the pressures are the greatest. These piping failures occur with such regularity that the FDNY firefighting procedures established alternative water supply operations where the pumper supply line is connected to the first floor standpipe outlet in the event the siamese connection piping failed. Our new building code has eliminated this high-pressure failure problem by requiring a separate siamese piping system that can withstand the high pressures necessary when supplying water to upper floors. Another goal was to eliminate pressure-reducing valves (PRV) from standpipe outlets. PRVs are a nightmare for the fire service.

Fire Chronicle

The FDNY firefighting procedures, highrise buildings SOP for standpipe operations requires that the PRV be removed from the standpipe outlet to eliminate any restrictions in water flow.[34] Over the course of many years the installed PRV coupling binds to the standpipe male outlet, making it very difficult to remove.

During standpipe firefighting operations, a firefighter is positioned at the standpipe outlet to control pressures by opening or closing the outlet valve. As a firefighter assigned to this position, I have struggled to remove these couplings from standpipe outlets set up close against a wall while the fire attack team was calling for water in the line. I can attest to the fact that this PRV removal process definitely increases the amount of time it takes to get water on a fire and has contributed to many pulled muscles and bloody knuckles.

* * * *

In the new code, these devices were eliminated from the standpipe outlets by having 300-ft zones with each zone installing a PRV in the down-feed water supply pipe, not on the outlets, to reduce operating pressures. No PRV removal at the standpipe outlet is necessary, something I know firefighters will appreciate in new building construction. NFPA 14 provides the accepted standard for minimum standpipe requirements.[35] Our modification of this standard went beyond the "minimum" requirements and had to be justified. When construction professionals and building owners asked the question, "Why isn't the NFPA standard good enough for New York?" we had to provide an answer.

Safety costs money. Every fire protection feature with a history of providing a high level of fire protection—sprinklers, fire tower stairways, cement-encased structural steel, fire-resistant compartmentation, structural redundancy—costs more than the "minimum" alternatives. In effect, many NFPA and other construction industry standards become the code because most fire departments cannot obtain the political support needed to change building codes that increase construction cost because of safety features.

It is for this reason that I am sharing some of the design information of the standpipe system that the FDNY developed for the NYC building codes. It is a ready response to the question of why to spend more. By working with, not against, building construction professionals, we were able to design a redundant water supply and piping system that increases the reliability of both standpipe and sprinkler systems.

We took a lot of criticism over these changes that added cost and lost rental space in a building due to the additional water storage. We were able to prevail by reviewing existing NYC codes to find out what worked and what should be changed. One item that we felt could be changed was the requirement for a manual pump installed in the building with the capabilities of supplying water to all floors and required a building engineer on duty 24/7 to operate it. In exchange for eliminating this manual pump, we increased the standpipe water supply for every 300-ft zone above the initial zone and required a dedicated standpipe riser for buildings over 600 ft.

This standpipe system gives firefighters a fighting chance to prevent fire-induced building collapse. And it should be noted that the water in these gravity tanks can also supply the sprinkler systems, providing adequate redundancy and increasing the effectiveness for these systems as well.

The following is an overview of the general features if this system:

- Each required exit stairwell has to have a standpipe riser.
- The reserve water system must supply the required flow rate at a minimum of 65 psi to all outlets.
- The system must supply the flow rate for minimum duration of 30 min (this is based on the type of occupancy and can be more).
- Tank supplies must be divided into at least two individual tanks or compartments of equal capacity.
- Minimum riser diameter will be 6 in.
- Systems will be separated vertically into zones of no more than 300 ft in height.
- Each zone will have its own individual water supplies.
- Zones above 300 ft must have a gravity tank supply.
- Zones over 500 ft must be "dual-fed" (down fed by a gravity tank and up fed by a fire pump).

For buildings over 600 ft, additional features apply to provide a secure redundant water supply:

- Each zone will be up fed by an automatic fire pump. The fire pump for any upper zone may take its suction supply from the gravity tank supplying the lower zone. The pump for the lowest zone may take its supply from city water mains.
- Separate fire department connections will be provided for every 600 ft vertical of the building (i.e., a fire department siamese connection will be provided for the lower 600 ft of the building, a separate fire department siamese connection, with piping rated to accommodate high pressures, will serve the next 600 ft via an express riser).
- Maximum fire department-supplied pressure, except for the express riser, will be 350 psi.
- Maximum fire department-supplied pressure on the express riser will depend on building height (e.g., 900 ft will be approximately 480 psi, 1,200 ft will be approximately 600 psi).

This system ensures that the upper floors of a "super highrise building" (greater than 600 ft) will have a sufficient redundant water supply provided by the gravity tank in the zone, plus the gravity tanks in the zone below, and a separate standpipe riser that can handle the high pressures necessary when water is supplied by a fire department apparatus.

The standpipe siamese feeds water directly into standpipe system, not into gravity tanks. This adds the additional source of the domestic water supply in providing gravity tank water.

These means of supplying standpipe water to the upper floors of a highrise building gives adequate support and an increased level of safety to firefighters operating in this environment (fig. 4–4).

The full details of this standpipe system can be found in the NYC Building Code, Chapter 9, Section 905, Appendix Q (NFPA 14 as modified).

In order to ensure that the FDNY pumpers could provide an effective hose stream on the upper floors of a highrise building, I conducted a drill on high-rise pumping procedures where we pumped water from a pumper connected to the standpipe siamese up 729 ft to the roof level. We then stretched three lengths of 2½-in. hose line at the roof level with an open tip nozzle off the standpipe outlet. Once 400 psi was reached at the pumper outlet gauge, the standpipe outlet was opened and the line charged. While the line operated for 15 minutes, the standpipe outlet gauge indicated 65 psi, the nozzle gauge 40 psi, and the gpm (gallons per minute) gauge at the pumper outlet reflected 230 gpm flowing. This is an acceptable stream for interior firefighting operations.[36] The drill demonstrated that no matter what happens to the building's pumping systems or water supply, the FDNY can provide an effective hose stream. It should be noted that this drill was conducted in the newly constructed 7 World Trade Center building and it was the first time the FDNY had used high pressure pumping in a drill.

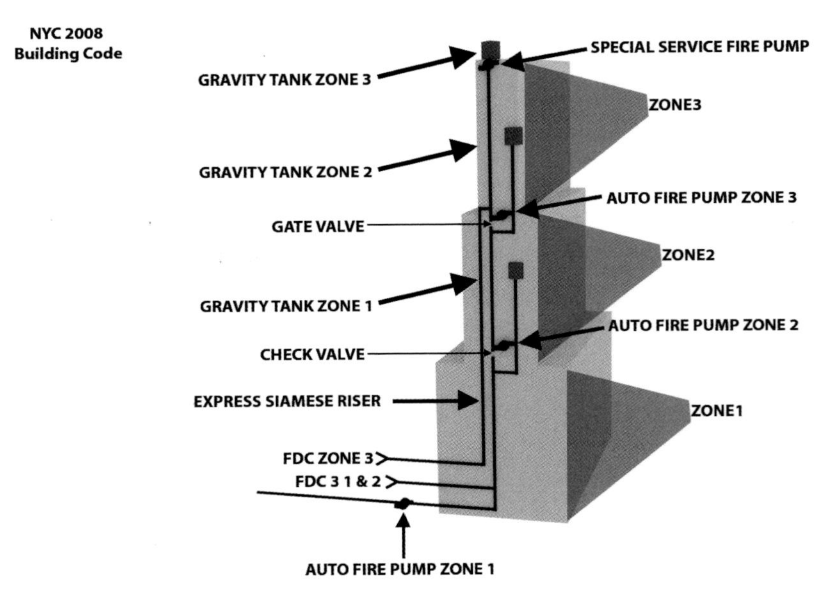

Fig. 4–4. Minimum standpipe requirements

There were a several unexpected events during the drill. The "wet" standpipe system should have had its piping filled with water, but as the system was being charged the gpm gauge at the pumper outlet indicated a flow of water. It was discovered that water was filling the standpipe riser pipes in some sections of the piping that lacked water. Once the entire system was filled with water, the gauge dropped down to 0 gpm. Another unexpected event involved the building's alarm panel. Several water flow trip valves on different floors activated as the system was being charged. The increase in air pressure caused by the water being introduced into the standpipe riser triggered the alarms. Once the building engineer reset the alarm devices, they stabilized and returned to the normal setting.

Every fire department that has highrise buildings where high pumping pressures are required should conduct high-pressure pumping drills to ensure that the system is working and that fire department pumpers can provide the required pumping pressures for hand line operations.

Compartmentation and expected areas of extinguishment

The fire area that a professional team of firefighters can be expected to extinguish needs to be determined. Some say that for fires in a typical highrise office building, a 2,500-sq-ft fire area is the maximum size fire that can be extinguished by firefighters using a 2½-in. handheld hose line with an open tip nozzle off a standpipe outlet providing a 50-ft hose stream reach.[37]

When NYC created Local Law 5 for highrise buildings it required compartmentation of areas over 7,500 sq ft. This was probably based on the standpipe being designed to deliver 750 gpm for 30 min (three 250 gpm hand lines).

The ability of manual fire suppression to successfully confine a fire to 7,500 sq ft was challenged in a report on the NYC City Bank Building fire:

> Interior attack was incapable of controlling the fire on the sixth floor and could not prevent extension to the seventh and eighth floors. The exterior elevated streams that did control the fire were effective at the sixth and seventh floor levels and could have been used up to the ninth or tenth floor, but on higher floors they would not have been capable of gaining control of the fire . . . . The assumptions that a combination of passive structural fire protection and manual fire suppression will successfully confine a fire to 7,500 or 15,000 square feet appear to be overly optimistic, judging from this experience.[38]

The only support for the 2,500-sq-ft extinguishment calculation that I found was a study performed in the 1950s which resulted in the Iowa rate-of-flow formula. This formula, cubic feet/100 = gpm, defines the rate of flow needed to control a fire in a single open area when that area is fully involved in fire. The formula states that the cubic feet involved, divided by 100, equals the gallons per minute necessary to knock down a fire in an area if that flow is properly distributed over the entire area. The formula is based on and supported by the heat absorption capability of water, the heat production based on the volume of air in a given open area, the steam generated and the amount required to displace air in a given area.[39] This formula predicts 250 gpm is needed for a 2,500-sq-ft compartment with 10-ft ceilings.

A handheld 2½-in. hose line with an open tip nozzle is capable of delivering up to 250 gpm of water. This, however, is an optimum volume. The back pressure and need to maneuver the line often requires partially shutting down the nozzle to gain control, which results in decreasing the volume of water being delivered on the fire.

Tools are being developed that are capable of delivering high volumes of water inside a building, such as a portable large caliber stream appliance that is capable of delivering in excess of 250 gpm. When a portable large caliber stream appliance is in position it can continually operate at full capacity, provide greater stream reach and gpm.

When fire chiefs were surveyed about the 2,500-sq-ft expected area of extinguishment by firefighting personnel using the 50-ft reach of a handheld hose stream, their estimates were both larger and smaller than a 50 × 50-ft area.

I have operated at fires in highrise buildings for more than 34 years and have found that the area of expected extinguishment can vary. The size of the area depends on the variables involved: type of material burning, building construction, air movement, and standpipe outlet water pressure. All of these variables must be factored in and tested under controlled conditions to determine the expected extinguishment area accurately. Hopefully, a respected agency such as the NIST or the National Fire Academy will perform the testing necessary to verify this formula.

Once an up-to-date scientific test is performed that accurately determines the expected extinguishment area, laws concerning compartmentation can be redefined based on a more definitive concept of the level of fire protection that hose stream operations can provide. For now, the current acceptance of the 7,500-ft limit sets the parameter for compartmentation.

Any fire department using less than a 2½-in. diameter hose line off a standpipe outlet to fight a highrise building fire should consider the decrease in the amount of water flow versus the gains of carrying the lighter equipment. A 1¾-in. diameter hose line generally will flow water at the rate of 150 to 180 gpm compared to the 250 gpm flow of a 2½-in. hose line. Using a smaller diameter hose line decreases the area of expected extinguishment to a dangerous level and offers less protection to personnel operating in the area. The key factors in successful highrise fire operations that will reduce the incidents of fire-induced building collapse requires the use of the incident command system, a prefire plan, the proper manpower and equipment, and a building system that supports these efforts.

Firefighter Fatality Reports

Reviewing reports where fire-damaged collapsing structures caused a firefighter fatality can provide insight into first responder actions as the situation progressed and ways to recognize potential structural collapse warning signs to protect operating personnel.

The following firefighter fatality investigations highlight the importance of risk management in determining fireground operations when the stability of a burning building is in doubt.

In order to understand how to apply the ideas brought out in this book, after each investigation report an attempt should be made to critique the incident using risk management. I have followed the investigations with my own risk management analysis, which can be used as a benchmark to judge conclusions.

Firefighter fatality investigation 1

On February 14, 2000, in Houston, Texas, the vulnerability of firefighters operating in a fire involving a truss roof was once again demonstrated when two career firefighters died as a result of the sudden collapse of a truss roof in a McDonald's restaurant.

A fatal fire investigation reported the roof system as lightweight wood trusses, consisting of 2 × 4 and 2 × 6-in. lumber connected with metal gusset plates. The total span of the trusses over the majority of the structure was 47 ft, 6½ in. There were no interior load-bearing walls noted. The trusses were placed 2 ft on center with 2 × 8-in. lateral bracing. The exterior roof system consisted of 5/8-in. plywood sheathing with fiberglass asphalt built-up roofing. There were a total of five HVAC units located on the roof (three 10-ton units and two 5-ton units). The interior ceiling of the building was suspended 1 ft from the trusses, using suspended ceiling panels.[40]

It was not determined exactly when the truss roof collapsed. The fire was first noticed at 04:38 hours by Medic 73. They reported to dispatch that they had visible fire emitting through the roof. The first fire suppression operating units arrived one minute later.

At 04:51 hours there were approximately 30-ft flames extending from the center section of the building and the incident commander could see heavy fire in the kitchen area. Based on this observation of the fire conditions, at 04:52 hours the incident commander decided to evacuate firefighters from the interior and order a defensive attack. Shortly after the command to evacuate the building, the incident commander ordered a "personal accountability request" from all companies on the scene. This procedure determined that two members were not accounted for. Within approximately 14 minutes of fire suppression operations, the roof structure failed.[41] Based on information obtained from the City Fire and Arson Bureau, it was determined that the fire started in the office area approximately 25 min prior to Medic 73's arrival, which would put the actual burn time from start of fire to the evacuation of interior fire suppression forces at approximately 39 min.[42]

Risk management analysis. While the exact cause of the collapse was not determined, the amount of fire showing through the roof indicates that the fire involved the roof's structural members. It has been well documented that when a truss roof structure is involved in fire, it may be expected to collapse within 10 min.[43] This is definitely a "high severity" collapse situation and should be evaluated as such. When only one truss element is weakened by fire damage, the entire roof system becomes prone to a progressive collapse.

Several stress/load factors of the roof construction to be considered were an unsupported span of 47 ft and the HVAC units putting the truss elements under significant tension. The dropped ceiling fire protection failed to protect the truss structural elements from fire damage.

Another feature of this building that should be part of the risk evaluation is that the existing building code did not require sprinklers in this type of occupancy and the building was not sprinklered.

Regarding this collapse, what can be determined is that the lack of effective passive structural fire protection (dropped ceiling), no immediate active fire protection (sprinklers), combined with the fire damage to the lightweight truss structural elements created the conditions for a rapid progressive collapse, which resulted in a catastrophic roof failure.

The risk analysis should include whether or not sprinklers are present. Sprinklers are a factor in assessing the amount of fire damage to the structure. If sprinklers are not present, total structural fire protection is provided only by the passive fire protection methods used. It can be concluded that when sprinklers are not actively operating, the structural members are prone to suffer more fire damage during the burn time of the fire than when the sprinklers are working to control a content fire.

The presence of sprinklers should not change conclusions based on time (how long has the structure been burning), type of construction, and intensity of the fire. These considerations are a mandatory part of the risk analysis when determining the potential for building collapse. In order to make risk analysis routine for a incident commanders certain actions must be performed as a matter of habit. At the moment when the alarm comes in, the IC should start thinking about risk identification: What type of structure/occupancy is involved? Does a prefire plan exist for the building? How does the time of day impact conditions? Is there a normal response of units to the scene? While responding, a risk evaluation of the first-arriving unit's radio reports should be performed. Upon arrival and after initial risk identification, the thought process should shift to risk priorities and risk control. Using this process will assist in making the evaluation routine.

A report sponsored by the National Fallen Firefighters Foundation and the U.S. Fire Administration highlighted the need for risk management:

> Focus greater attention on the integration of risk management with incident management at all levels, including strategic, tactical, and planning responsibilities.[44]

As stated at the beginning of this section, investigations in firefighter fatal fires is a good source of understanding how buildings collapse from fire damage. However, the leaders of fire service must sometimes take a hard look at what is being recommended before using any firefighter fatality investigation report's recommended procedures. Fire departments must use caution and thoughtfully consider the source of the report, what resources were provided to make the report, how the report is researched, and the specific recommendations when tactical operations are under review. (This should also be kept in mind when performing practice risk management analysis. Keep it real.)

While generally there is great value in the National Institute for Occupational Safety and Health (NIOSH) reviews of firefighting operations, the NIOSH investigation of the Pittsburg fatal church fire reveals a concern that federal agencies are beginning to establish firefighting operational procedures that can hinder and might prevent fire officers from doing their jobs effectively.

A fair understanding of the impact of the "recommendations" made by NIOSH was reported in a news article regarding firefighter fatalities:

> As a matter of policy, NIOSH does not criticize firefighting tactics in its reports, but makes recommendations on how similar future fires ought to be fought. Its recommendations, however, often amount to veiled criticism.[45]

It is accepted that NIOSH has good intentions to prevent firefighter fatalities and often makes sound judgments in their reports. The concern is when this is not the case and recommendations do not represent practical tactical field procedures.

Firefighter fatality investigation 2

A critique of the NIOSH investigation and recommendations for the Pittsburg, Pennsylvania church fire follows:

> The alarm was received 08:46 hours on March 13, 2004, a cold morning with winds gusting to 22 mph. The fire building was a church built in

> 1875, brick construction covered with stone, 120 × 70 ft, with a heavy timber roof truss.
>
> The bell tower, constructed of brick with a stone façade, was supported by steel I-beams connected into a corner of the original church. The basement had meeting rooms and an electrical room. There was also a small subbasement. The church had an annex, 60 × 45 ft, that was attached on one side. There were interconnecting hallways on all floors of the annex and a central elevator shaft.
>
> First-arriving units reported heavy smoke in the church area with a possible fire in the basement. At 09:28 hours, an explosive backdraft occurred. Six firefighters were injured by the blast and several were burnt when the fire rolled over the basement ceiling. Heavy black smoke filled the structure until 10:09, when visible fire broke through the roof.
>
> At 11:48 hours all exterior hose streams were shut down. Twenty-five minutes later the church bell tower collapsed, fatally trapping two firefighters who were overhauling inside the front entrance vestibule.
>
> On the exterior, falling masonry seriously injured the incident commander, who had to be removed from the scene, and several other firefighters.[46]

The NISOH report made 10 recommendations/discussions. These include the following:

NISOH Recommendation 1. "The IC should seek the help of qualified structural experts or other competent persons to assess the need for removal of weakened construction or make provisions for shoring up walls, floors, roofs, or as in this case, the bell tower."[47]

It is appropriate to routinely call in engineers to assist in providing information on the stability of a building suffering significant fire damage. However, fire officers' judgments concerning the fire building's stability should not be underestimated. This was demonstrated previously where the building department's engineers stated it was safe to enter the fire-damaged building, but the chief officer felt that the amount of water damage from high-caliber streams used for hours had undermined the building in ways that were not obvious to the building engineers. His judgment was confirmed when the buildings suddenly collapsed shortly after being declared safe by the engineers. The chief's knowledge, while not infallible, often does have value when the structural expert engineer has little or no experience evaluating fire-weakened structures.

From reading the reports and researching the available information, an understanding of the collapse of the bell tower is as follows:

- The bell tower collapse was sudden. It occurred without warning, which indicates that there were no reported structural defects visible to operating members (no cracks or structural distortions) or any falling debris reported prior to the imminent collapse.
- The structural failure of the bell tower can be attributed to several factors. Primarily, significant fire damage occurred; the fire through the roof and an earlier reported flash over involving much of the cellar indicated that this fire went beyond a content fire and was actively weakening the structural members.
- The fire damage and added load of weight from the water used in extinguishing the fire created conditions of nondesigned fire damage load stresses. These stresses, combined with the exterior stone veneer load, which had been standing since a 1930s renovation and naturally would have produced some distortion to the supporting steel I-beams, most likely resulted in the catastrophic effects on the structural steel supports.

However, without structural failure data to support this assumption (it is possible that there were other causes such as faulty workmanship or additional alterations not identified), it is not possible to state with certainty that these forces were the primary cause of the structural failure.

NIOSH recommendation 2. "A defensive strategy was established within one hour after initial interior firefighting and at this time there should have been a collapse zone established, 1½ times the height of the structure. In this case the bell tower was 115 ft high, so the collapse zone should have been 173 ft away from the church in all directions."

NIOSH recommendation 3. "Fire departments should ensure that the incident commander establishes the command post outside of the collapse zone. The command post from which the IC manages the fireground must be located in an area outside of the collapse zone."[48]

These two recommendations would often remove a command post from any possible visual viewpoint of the fire, reducing immediate situational awareness and might hinder radio communications which would additionally impair command judgments.

To my knowledge, this is the first time that a collapse zone has been defined as 1½ times the height of the fire building. In my experience, it has always been considered to be at least the height of the building.

For almost all incidents the initial incident commander should have a command post located where a visual assessment of the conditions is possible.

Basic tactical procedures require an initial command post to be established in close proximity to field operations and, if conditions warrant, move the overall command post to a relatively safe location with an operations post in close proximity to ongoing firefighting activities.

The need for an incident commander to visually supervise complex operations cannot be underestimated. It provides situational awareness, which is essential to control operations safely and effectively.

NIOSH recommendation 4. NOISH recommendation 4 commented on the need to mandate training firefighters to recognize backdraft conditions, which included "confined fire with excessive heat, smoke stained windows, and little visible flame from the exterior." It went on to state that

> If any of these warning signs are present, firefighters should immediately relay the information to the IC, move away from doors and windows, and immediately exit the structure.[49]

The dynamics of actual fire conditions must not be underestimated. During my years as a fire officer if I had to immediately remove firefighters from a burning building because the fire building had smoke-stained windows or confined fires with excessive heat, there definitely would have been situations where fire would have claimed a higher toll of fire victims that could have been saved. Mandated restrictive "recommendations" by federal agencies can create conditions that put civilians and firefighters in more danger. If the fire is aggressively attacked when it is small, more lives are saved, including firefighters.

The chief in charge must be given the latitude to determine the risk versus benefit analysis and adopt tactics as necessary.

The families of the firefighters who lost their lives at the fire read the NIOSH report and takes what it says as the truth. The nation's firefighters deserve reports from federal agencies that live up to these expectations.

Federal agencies like NIOSH must be held accountable for their reports and conclusions. It's all about the ability to judge others' actions accurately and appropriately.

NIOSH recommendation 6. "Preplans should be conducted by first-due companies. Building construction could have been determined with a pre-incident inspection on church occupancies."[50]

If a Federal agency is going to define the standards by which a small town or big city chief is going to be judged, shouldn't it at least provide some support and help in meeting those standards? Real help like funding pre-incident planning. I believe it is unjust to set a government's standard that is financially near impossible to accomplish for many fire departments. NIOSH is basically a consultant company, advising the government on management and operational procedures. There are good and bad reports from consulting companies in the private sector and their recommendations are not always productive in the way intended. The same holds true of government agencies.

Certainly there is a place for these types of investigations, and I have lauded the National Institute of Standards and Technology (NIST) and NIOSH for their efforts in fatal firefighter investigations. Agencies that are charged with providing "firefighter safety" at the very least must have committee members with sufficient resources and time to conduct investigations. Their members must also have a widespread background in making fireground operational judgments to prevent injustices against the company and chief fire officers of this country.

We all need to be held accountable; we live in a time where this is a given. Mistakes should not be hidden, especially when mistakes involve the death of a firefighter. All firefighters of this country, which includes my family and friends, absolutely, positively, definitely must have good investigations when things go wrong and the best possible efforts made to prevent mistakes from happening again. The point I am making is that there has to be vigorous research by people who have extensive field experience in commanding firefighting operations before judgments are made and inappropriate, "mandated" procedures are passed for those in the "arena."

Firefighting is a profession that relies on practical knowledge. The experience of facing a building heavily involved in fire with occupants trapped is a type of situation that defines a firefighter's work and cannot be adequately described with words. It must be lived.

Firefighting is consistently reported as one of the most stressful occupations and the reasons for that have to be recognized. I can't help reflecting on how President Theodore Roosevelt put it:

> It is not the critic who counts: not the man who points out how the strongman stumbles or where the doer of deeds could have done better. The credit belongs to the man who is actually in the arena, whose face is marred by dust and sweat and blood, who strives valiantly, who errs and comes up short again and again, because there is no effort without error or shortcoming, but who knows the great enthusiasms, the great devotions, who spends himself for a worthy cause; who, at the best, knows, in the end, the triumph of high achievement, and who, at the worst, if he fails, at least he fails while daring greatly, so that his place shall never be with those cold and timid souls who knew neither victory nor defeat.[51]

NIOSH has a creditable reputation reporting on firefighter fatalities and other occupational hazards, and in most cases its recommendations provide valuable information. In many ways they are the voice of workers in dangerous occupations who need a forum where their concerns can be heard. BUT, NIOSH is not above criticism when it's necessary to help provide clarity to what firefighters require from a federal agency charged with providing this service.

Firefighter fatality investigation 3

The material for this report on a fire that resulted in the death of four Seattle career firefighters was primarily gathered from a well-balanced, thorough investigation.[52]

The fire building was primarily constructed of heavy timber members. It was constructed in stages and had been altered several times over its history of more than 85 years. The layout of interior occupancies was very difficult to understand from the exterior.

The streets in this area had been raised by 10 to 20 ft, making the "original" ground floor level, that housed a storage room and bakery, a windowless basement accessible only through the north and west wings. When the elevation of the streets was raised, an upper story was added with a new floor at the same level as the sidewalk. The new story was post and beam construction, with large-dimension wood members supporting the new floor and roof. At the time of the fire, the floor was constructed of a concrete topping over the wood floor to meet health standards for a food preparation facility. The new floor joists, of the now ground floor level along the north wall, were 5 ft higher than the old roof. To support the ends of the new joists, a wood frame "pony wall" (a low height wall) was fabricated from 2 × 4-in. wood members and erected on top of the older original building's wall ledge.

The original one-story building was now the "basement" beneath this new floor. The original, lower-level building was occupied by a bakery and storage room used by the Mary Pang Company. The main access to the storage room was through a sliding fire door from the loading dock in the bakery area.

This storage room was 30 ft wide, 60 ft deep, approximately 20 ft in height, and was heavily loaded with combustible products. The underside of the floor deck and the structure that supported it were exposed and unprotected, including the "pony wall." The room had no windows, only two doors, and three brick walls. The partition between the storage room and the bakery was wood construction.

The integrity of the entire upper structural system depended on gravity. There were no mechanical fasteners connecting the column caps to the floor beams or to the upper posts. The weight of the structure held all of the components together. Several of these connections failed later in the fire as the integrity of the entire structure deteriorated and loads shifted.[53]

Fire conditions on arrival. The arson fire was set in the storage room of the building shortly before 19:00 hours on Thursday evening, January 5, 1995. Engine 10 was the first-arriving unit. The lieutenant saw fire involving a shed on the roof of the building and reported a "well-involved building fire" at 19:07.

It appeared to Engine 10 that the main body of fire was coming from the small shed (this shed was used as a lunch room area for the workers) on the roof of the west wing of an "L" shaped building. They believed that the fire originated in this shed and was extending to the main structure. What they did not know was that the fire had quickly spread from the storage room basement, up into the ceiling, extending through the underside of the roof and igniting the shed on the roof. Ladder 1 joined Engine 10 on the east side, first floor area, which was the front of the new building, forced entry, and raised portable ladders to the roof.

Battalion 1 arrived one minute later and approved the interior attack plan to keep what they thought was a roof structure fire from extending into the building.

As companies began arriving, Engine 2 was assigned to look for additional areas where the fire could extend to the north and west areas of the building. They proceeded down to the lower ground level and located an opening in the north wall of the west wing that provided access to the interior loading dock. They advanced their line through this opening when they found the sliding fire door leading into the storage room partially open. The interior of the storage room was fully involved in fire; however, no smoke or flames were coming out of the doorway. The officer had been briefed on the attack plan, emphasizing that the objective for units on the west side was to prevent extension while the interior attack was made from east to west. They took a position to hold the fire at the loading dock, in anticipation of an interior attack coming toward them.

Floor collapse. At approximately 19:30 (23 min into fireground operations), three interior lines were being operated by Engines 10 and 13 and Ladder 7. Firefighters working on this floor believed that they had already gained control of the situation and were not aware that the main body of fire directly below them under the concrete floor, which was preventing the heavy fire conditions from being detected.

The Lieutenant and three firefighters from Engine 13 were operating in the northwest corner and the lunchroom. The lieutenant and one firefighter from Engine 10 were working in the north half of the ground floor, while the lieutenant and three firefighters from Ladder 7 were near the middle of the large space. All of these firefighters were in dense smoke, but the atmosphere was cool and there was no visible fire. The lieutenant of Engine 10 and his partner briefly encountered the crew of Ladder 7, and then disappeared back into the smoke.

Seconds later the building rumbled and flames erupted from the basement as the floor began to collapse. Sections of the wood and concrete floor hinged down into the basement. The flames coming from the basement spread across the underside of the roof and the contents of the ground floor began to ignite in a rapid flashover sequence. It was evident within seconds that something was going wrong. Personnel outside the building were not immediately sure what was happening. Hot, heavy smoke and some flames began to issue from the doors and out through the hole in the roof. The sudden floor collapse can be attributed to an unusual construction detail which would have been extremely difficult to anticipate or predict. The initial "size-up" of the building indicated a heavy timber building. The pony wall was an unprotected, small-dimension, wooden load-bearing member that was not designed for or capable of withstanding fire damage or any type additional loading stresses from the structural system. This unprotected wooden pony wall's 2 × 4-in. structural members collapsed due to fire damage that weakened its ability to support the floor joist load. The failure of this the pony wall initiated a progressive collapse, which caused the ends of the floor joists and the concrete floor to drop down into the basement.

Two firefighters from Engine 13 were in the lunchroom and heard their lieutenant shout "Let's get out of here!" as the floor began to drop. The two firefighters were able to go out through the hole in the wall and onto the roof of the west wing. The lieutenant of Engine 13 and the other firefighters are believed to have fallen into the basement as one of the first sections dropped.

The lieutenant of Engine 10 and his partner became separated as the floor collapsed. The firefighter was able to make his way back to the door and out, while the lieutenant dropped into the basement.

The members of Ladder 7 at first believed that the ceiling must have collapsed as flames spread across the open area over their heads. The lieutenant opened the nozzle, attempting to cool the overhead, but the water stream immediately turned to steam. The four firefighters began to follow their line back toward the door in single file as intense heat radiated down on them. They did not realize that the floor was collapsing until they encountered flames coming up through a large opening, almost directly in their path back to the doorway.

Two of the firefighters and their lieutenant managed to scramble to the door and outside, passing within feet of the opening. When they reached the exterior they realized that the third firefighter was no longer with them. The missing firefighter had been first in the line as they were trying to find their way out and he either fell into a hole or dropped into the basement as a floor section collapsed under him. All of the seven firefighters who escaped were burned.[54]

Review of risk analysis performed by operating members. The initial risk assessment, identified by the first-arriving officer and approved by the acting deputy chief, was an aggressive interior attack to keep the fire from extending into the main building. This east-to-west attack plan was effectively carried out. Companies were assigned to the west side to prevent extension of the fire and were specifically directed to avoid opposing hand line operations with the interior attack.

- Risk identification: The heavy timber fire building appeared to have fire extending from an outside small roof structure into the main building.
- Evaluate risk: Most significant risk was evaluated to be the fire extension.
- Establish priorities: It was decided to position interior hand lines immediately.
- Risk control: The three hand lines were operating on the fire to control fire extension.

Risk management analysis

Risk identification. The challenge to first-arriving fire officers involving complex fireground situations is to consider *all* the potential risk factors during the early stage of an incident. Easier said than done.

When there are large, complex building layouts, it takes time for the incident commander to gather all the related information on an unfamiliar building. And time is not on your side when there is an out-of-control fire involving a structure. If your fire suppression efforts are not actively diminishing fire conditions, you are losing the battle and conditions are getting progressively more hazardous. There is a very limited amount of time to reflect and make judgments.

The structure's exterior made it difficult to understand the interior arrangement or the construction details. The building's initial size-up from the exterior with the steep grade levels at the north side made it impossible to drive and difficult to walk all the way around the building. To make a complete recon of the building's layout required a significant amount of time by a qualified officer.

There were no prefire plans to assist in a risk identification. Even if there had been an inspection of the site for a prefire plan, it would have been difficult to discover the insubstantial and unprotected pony wall construction feature.

Risk evaluation. As this incident progressed, the attack plan appeared to be working. The visible fire was controlled and there were no reports of significant interior fire damage.

This investigation discovered that as the fire operations continued, there was a lack of progress reports. These progress reports would have revealed new information that did not coincide with observations from other points. If the incident commander is not getting pertinent information, the process of risk evaluation will not be effective.

The interior attack crews on the first floor did not report that very little fire had been found inside the building and all flames appeared to have been knocked down. The fire in the basement had been located, but it was not reported back to the incident commander. Engine Company 2 did not recognize that no one else was aware of the large interior fire or that the interior crews were directly over it. This was critical information that could have changed the risk analysis and caused an evacuation of the interior crews before the floor collapsed.[55]

During large, complex emergency conditions, where many units are engaged in fireground operations, it is common for an incident commander to not receive all the information of individual unit's activities. At the beginning of an incident, handie-talkie communications are often on only one channel, which can become overloaded with communications. This one channel must handle the necessary inter-unit communications as well as unit officer communications with the incident commander. The trend to equip all members with handie-talkies to increase the safety of each operating member has resulted in an increase in the volume of radio messages. Incident commanders at multiple alarm fires must deal with "information overload" from having numerous radio-equipped members on the scene and also the need to communicate with central command on existing conditions.

In the 1970s handie-talkies were used by a limited number of members and, while fires were the same, the amount of communication was far less than what it is today. This allowed the chief in command to spend more time focusing full attention on fireground operations. I have witnessed how initial radio communications have improved over the years and have also experienced the difficulty in handling multiple radio channel frequency messages. Sometimes, during intense fireground activity, even monitoring only two channels while actively engaged in supervising multiple operations is almost impossible.

Today, the communication demands on the incident commander have increased dramatically. Technology now permits real-time communications over a wide range of devices and *everyone* wants to know what's going on: police departments, emergency management commissioners, press office, news media, and so on. This phenomenon should not be underestimated. An incident commander now works in a more congested communication environment that can often distract from the focus necessary to get the big picture of the entire fireground. In fact, communication overload can be an impediment to understanding what is going on at the fire scene.

This investigation recognized this issue and made these recommendations:

> A progress report from either of the two division supervisors could have caused the other division supervisor to recognize the inconsistency and alert the incident commander to the problem. The lack of progress reports appears to be related to the lack of aides or assistants to support the division supervisors. When division supervisors cannot directly monitor the work that is being performed by their assigned companies and the fire conditions in their assigned areas, there is a risk that several important functions, including supervision, coordination, safety surveillance, and progress reporting will be compromised.

> The division supervisors are responsible for these functions as well as for monitoring two separate radios, one on the tactical channel and one on the emergency channel. These important responsibilities tend to keep the division supervisors outside, at secondary command posts. They must rely on company officers to perform the interior supervision and coordination functions and to keep the division supervisors informed about what is happening inside. Some of these responsibilities could be delegated. However, the battalion chiefs, who are normally assigned as division supervisors, do not have aides. Also, because of the rule that a minimum of two personnel are required to make an interior entry, a division supervisor would need a partner to be able to go inside. To fulfill all of the requirements and allow the division supervisor to enter the structure, two assistants would have to be assigned to each battalion chief. This would require a company or other available personnel to be assigned to support each division supervisor.[56]

I completely agree with these conclusions. Often when I was initially engaged in a large, complex, heavy fire operation, communications became overwhelming at some point and it required communication assistants and aggressive actions to control handie-talkie communications in order to get a clear understanding of the immediate conditions. The only solution to prevent these noted shortcomings is to have a trained aide and a field communications company assigned to assist in monitoring multiple communication frequencies, relay communications to central command, and act as a buffer for other agencies' communication needs and the news media.

What should not be lost in all these communication features is that all firefighters are responsible to continually observe conditions and immediately inform their superiors or the incident commander if they become aware of any information that could impact the overall strategy for the incident (risk identification and evaluation). In order to avoid "information overload" and tie up the operation's radio frequency, training must ensure that all messages use brief, standard, clear terminology and that emergency information is transmitted without delay in a recognized manner (risk control, the use of urgent or mayday transmissions).

This report made the following pertinent observations about the incident commander's responsibilities while performing risk identification and evaluation:

> This emphasizes the value of a complete, 360-degree size-up of the fire scene as early as possible by the incident commander or by an individual who can report in person to the incident commander with a "full picture" of the scene. In the absence of personal observations or prefire plans, the incident commander must rely on others to provide information from their different vantage points, then must assemble and interpret their observations. . . . The comparison of indicators from different vantage points often reveals significant information about a fire. It is important to look at the "big picture" as well as individual, narrow views.[57]

An incident commander must make critical life-and-death decisions based on experience, a visual assessment, and radio reports defining a fire building's structural features and the amount of damage structural members have sustained. The risk management system provides a framework to help accomplish this extremely difficult task:

> At this fire there were two senior chief officers with the experience to recognize inconsistencies between fire conditions inside the building and on the exterior. If they had the time to discover the basement fire it is probable that immediate actions would have been taken to address this risk.

Unfortunately, the fire did not provide that time. The investigation concludes with some observations that should be planned for:

> Situations that require simultaneous fire cause and line-of-duty death investigations are complicated and require a high level of coordination among the responsible individuals. The complexity increases rapidly when criminal acts are involved and additional organizations and agencies become involved in the investigation process. Good preparation and planning, along with positive, established relationships, are essential in these situations.[58]

A sound recommendation that can result in a better prepared fire department. All in all, this well-balanced, credible fatal fire investigation provided valuable information to the nation's fire service. There have been a number of situations where firefighters have been operating inside a fire building on a floor level that appeared to be safe, not knowing that they were directly above a serious basement or lower level fire and these conditions resulted in sudden and unanticipated floor collapses, which either dropped the firefighters into the fire area or exposed them to an eruption of fire from the level below them. While these types of incidents are rare, they do occur.

In fact, this type of event was recorded in New York City.

Firefighter fatality investigation 4

The fatal fire events on October 17, 1966, began at 9:36 p.m. when the bells sounded in Manhattan's firehouses and the house watchmen turned out their companies for Box 598. Engines 14, 3, 16; Ladder Companies 3 and 12; Battalion 6; and Division 3 were the first to respond.[59]

They were confronted with a serious cellar fire at 7 East 22nd Street. A heavy volume of smoke and heat were pushing from an art dealer's cellar and first floor of the building. Deputy Chief Reilly transmitted, "Using all hands" at 9:58 p.m., which brought Rescue 1, Battalion 7, and the Field Communication unit.

Battalion Chief Fredrick White determined that immediate extinguishment of the cellar fire was not possible and attempted to control the first floor to prevent extension to the upper floors. As heat conditions intensified in the cellar, he withdrew the units from that area and concentrated the efforts on the control of the first and higher floors, while making provisions for the examination of the exposures. He was handicapped by a lack of information regarding the building's construction, alterations, interconnections, and occupancies. The extensions in the rear of the buildings deprived him of access and seriously interfered with his reconnaissance.

Division 3 requested an additional truck company at 10:03. Ladder 7 turned out under command of Lt. Finley, a 29-year veteran of the department. Deputy Chief Reilly ordered a second alarm at 10:09. Engines 1, 5, 18, 33,

Search Light 21, the Super Pumper system, Safety Unit Car 500, Photo Unit Car 33, and Deputy Assistant Chief Goebel responded.

Deputy Chief Allen Hay of the 1st Division arrived at 10:09 and, after surveying the situation on 22nd Street, was directed by Chief Reilly to join Chief Higgins at 23rd Street.

By the time the second alarm was transmitted, fire had extended to exposure 2: an "L" shaped, three-story brick commercial structure at 940 Broadway. It was located on the northeast corner of Broadway and East 22nd Street.

In addition, fire had also spread to exposure 3, located at 6 East 23rd Street, a five-story, brick commercial building (fig. 4–5). The building on 23rd Street had three stores on the first floor: the Wonder Drug and Cosmetic store; FAN's, a women's dress shop; and Barton's candy store.

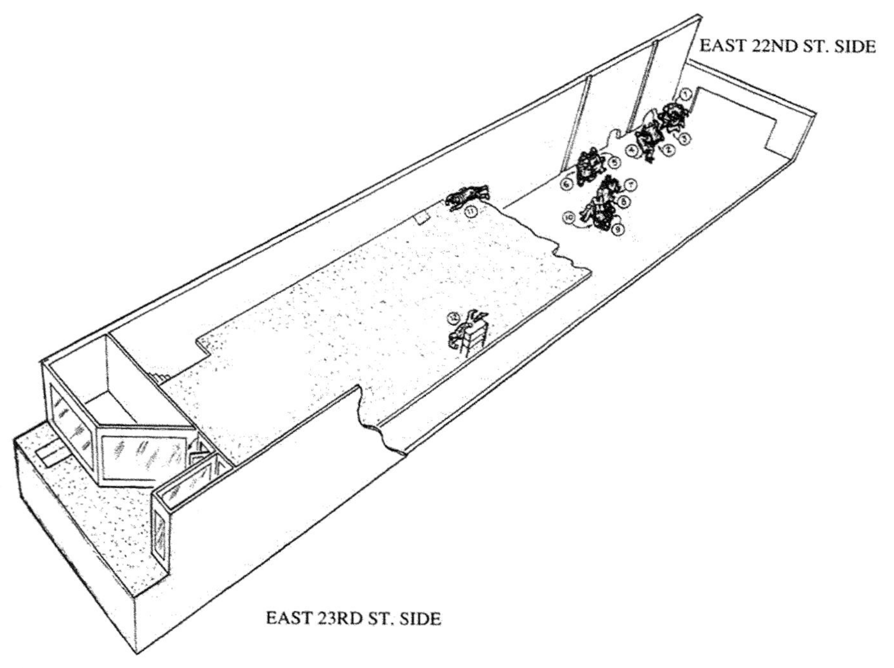

Fig. 4–5. 23rd Street fire

The fire had originated in the portion of the cellar occupied by the art dealer in the area partially beneath the Wonder Drug and Cosmetic store. The fire burned for a considerable length of time in this area and involved the upper portions of the stock and wood beams supporting the first floor northward to the cinder block wall, 65 ft from 23rd Street.

The timber floor beams were 3 × 14 in., covered with ¾-in. planking and mortared into the recesses in the brick wall. The planking was covered with 2 in. of mortar and a terrazzo cement floor tile surface. The 5-in. thick terrazzo floor acted as an insulator, which obstructed the transfer of heat and the flow of smoke and gases. The rigidity of the concrete floor masked any springiness normally accompanying gradual weakening of supporting beams.

Ladder 7 operated initially in heavy heat and smoke on the second floor of the original fire building at 7 East 22nd Street. After a short while they were ordered by Division 3 to "see what was showing in exposure 3."

Deputy Chief Reilly made a reconnaissance from the roof of 6 East 23rd Street and could see that fire had extended via a shaft between the buildings of 7 East 22nd Street and 940 Broadway, which was around the corner. He contacted Assistant Chief Harry Goebel and suggested that a third alarm be transmitted. Four companies—Engines 5 and 18 and Ladder companies 3 and 7—were directed to attack the fire from the interior of the 23rd Street building, whose rear wall abutted on the original fire building at 7 East 22nd Street. These companies entered the Wonder Drug and Cosmetic store, with Engine 5 and Ladder 3 descending into its cellar and Engine 18 and Ladder 7 operating on the first floor.

The companies on the first floor found no fire or heat and only light smoke. Engine 18 and Ladder 7 operated under the direction of Chief Higgins, who had sent his aide to the cellar in order to keep informed of the progress there.

Lt. Finley sent two members of Ladder 7 to the upper floors to search and ventilate there. They were joined by members of Ladder 12 to perform the same mission.

Firefighter Cicero of Engine 5, whose company was working in the cellar, was stationed at the head of the interior stairs at the front of the store by his captain to keep the company apprised of developments on the first floor. Almost immediately it was obvious the cellar was untenable as a heavy body of fire existed in the rear of that cellar. The officers ordered their men out, as flames were spreading along the ceiling above their heads. Lieutenant Royal Fox, in command of Ladder 3 (it was his first fire as an officer), remained in the cellar until certain all his firefighters were out safely. He carried the last firefighters up the stairs and both were taken to Bellevue Hospital for burns. He received the "William F. Conran" medal for his actions.

Meanwhile, Engine 18 and Ladder 7, under the direction of DC Thomas A. Reilly and BC Walter Higgins, were all unaware of the intense fire conditions that were directly beneath their position.

Chief Hay, shortly before, had returned to the street where he surveyed the overall situation and peered into the drugstore, which was clear enough for him to see the rear of the store.

Suddenly, without warning, a 25-ft section of the rear of the drugstore floor collapsed, hurling ten of these firefighters to their death in the inferno below. Two other firefighters trapped on the first floor by a burst of flame caused by the collapse were also burned to death.

Other than the chauffeur, the only other member of Engine 18 who survived that night was Firefighter John Donavan, who had been detailed to check parking violations at fire hydrants in the district. When he discovered that Engine 18 was operating at a second alarm he grabbed his turnout gear and drove to the scene in his car. He had just arrived in front of the Wonder Drug and Cosmetic store when the collapse occurred. Grabbing a line, he proceeded into the store and inched his way toward the rear of the first floor in heavy smoke. Suddenly he stepped into the void caused by the collapse. As he fell, he was able to clutch the nozzle with three fingers of his right hand. He hung suspended over the raging inferno below as his rubber coat started to ignite from the intense heat. All of a sudden, a hand dragged his collar and another his left arm and then his shoulders, and the two firefighters removed him to the sidewalk. These firefighters thought they were attacking the fire from the rear and had no reason to believe that substantial fire raged directly below their position. In fact it was thought that the main body of fire was still centered in the building to the south on East 22nd Street.

Attempts were made immediately after the collapse to rescue those trapped into the cellar, but flames had spread to all floors of 6 East 23rd Street, 7 East 22nd Street, and 940 Broadway. The rapidly spreading fire caused all of these buildings' roofs to collapse. Department members were ordered out of these building. Just as they backed out into the street, floor after floor of all three buildings crashed down until only the walls remained.

At 10:37 pm, about the time of this collapse, Car 17, Deputy Assistant Chief Goeble, ordered a third alarm. A fourth alarm was transmitted at 10:47 and a fifth alarm at 11:28 p.m. It was decided not to declare the fire under control until all the entombed men had been recovered.

Operations were hampered in the search by continual flare-ups within the still burning structure during the predawn hours as well as the present danger of additional collapse. Tons of debris had to be removed by hand in order to reach those still trapped below.

Throughout the morning of October18 and into that afternoon, the charred remains of these men were taken from the drugstore and placed in a waiting ambulance to be driven under police escort to the Bellevue Hospital morgue. The assembled throngs of grieving firemen removed their helmets and bowed their heads in silent tribute.

At approximately 1:45 pm the last body was removed and Chief of Department John T. O'Hagan gathered the firefighters in Madison Square Park and led them in an impromptu memorial service. He described that day as the saddest day in the 101-year history of the fire department and said, "I know we all died a little in there."

The rescuers were greatly aided in their search by a member of Fire Patrol 1, Edward Pospicil, who had observed some of the ill-fated men attacking a hot spot on the east side of the Wonder Drug and Cosmetic store just before the collapse. Shortly afterward, Patrolman Pospicil drew a map of the location where he had last seen the men operating. From this sketch it was surmised the when the floor collapsed, the men were pitched into the cellar at its southeastern corner. The bodies were found very close to the spot indicated on this map.

Like many of the members of the department who voluntarily responded on September 11, Carl Lee of Ladder 7 was on tag summons duty when the initial alarm came in and he voluntarily signed into the company journal to take the run in. He never returned.[60]

Review of fire prevention. Access to the cellar could only be had through an interior stairway in the Wonder Drug and Cosmetic store at 6 East 23rd Street. There was no cellar entrance in the street nor was there any interconnection on the first floor level of the building at 6 East 23rd Street with 7 East 22nd Street. The width of the cellar area occupied by Wonder Drug and Cosmetics was the width of the building. The length of the drugstore's cellar area had been diminished by the extension of the West 22nd Street art dealer's cellar, which had been built without application, permit, or building notice. The resulting situation created a deceptive picture, which would lead an inspector to believe that he or she had inspected the entire cellar of 6 East 23rd Street, when in fact an additional 35 ft deep area existed behind the rear, southernmost cinder block wall.

The cinder block wall separated the area of the drugstore's cellar from the area used and occupied by the art dealer. The drugstore occupied a store that was 100 ft deep and a cellar that was only 65 ft deep. The art dealer's cellar extended 135 ft from East 22nd Street, with the northernmost 35 ft beneath the drugstore. The 12 firefighters on the first floor were standing on a floor made up of 5 in. of concrete and were situated directly over the main body of fire.

On October 24, 1963, an inspector from the local fire prevention district office, while processing an application for a combustible permit for 2 East 23rd Street, and while in an area underneath the sidewalk on 23rd Street, discovered the passageway leading from 23rd Street to 7 East 22nd Street. Observing the employees assembling picture frames, he forwarded a memo to the local fire company requesting an evaluation of the need for a sprinkler system. The referral was not made on the customary form and was not properly identified and processed in the normal administrative work that exists in a company district that has the many varied occupancies such as Engine Company 14's. It was filed in the occupancy folder rather than scheduled for an inspection and the evaluation was never made.[61]

The Board of Inquiry made several recommendations:

- Compile computer data on all commercial buildings and make the information available to the chief in command of the fire.
- Participate in the Federal Fire Safety Act. (I have no information regarding the funding or policies advocated by this act, but maybe fire departments of today should resurrect it.)

- Develop an annual inspection program of commercial buildings previously profiled as potentially hazardous for firefighting purposes to assure that all legal requirements have been met.
- Establish formal coordination procedures between the fire department and building department inspection programs.[62]

The recommendations resulted in buildings identified as target hazards, which required a chief officer to inspect the structure, develop firefighting plans, and conduct fire drills. Also other steps were taken to provide this information at the fire scene for the officer in command of the fire.[63]

Many similar recommendations were made for the 9/11 WTC attack and the work continues to reach these goals to protect civilians and firefighters.

The 23rd Street fire is still remembered by members of the FDNY. At the Holy Name Society Annual Mass in May of 2006, the homily included a reference to the 23rd Street fire and the funeral mass held at St Patrick's for the 12 firefighters, which was also performed for members who died on 9/11.[64]

Risk management analysis. The similarities of the Pang Warehouse fire and the 23rd Street fire show how an unrecognized building feature contributed to a building collapse with fatal consequences. Risk identification is critical for a safe operation. Old buildings are often renovated to meet the demands of many different occupants. Whenever a serious fire involves these types of commercial buildings it becomes even more imperative to conduct risk identification with a conclusive survey of all areas of the building and its exposures to prepare for a sudden change in tactics as more information is gathered. These fires demonstrate that even well-disciplined, trained, and equipped fire departments will be faced with situations for which they cannot be completely prepared.

These two reports point the way to control risk by performing fire prevention inspections and having building codes that support the installation of sprinkler systems.

Firefighter fatality investigation 5

This investigation provides insight into how a series of events in a relatively small residential building can hinder an effective fireground risk analysis.

The fire occurred on April 4, 2008. A 37-year-old career captain and a 29-year-old part-time firefighter were fatally injured when a section of floor collapsed and trapped them in the basement during a fire at a residential structure.[65]

The fire structure was a two-story, single-family home of approximately 2,050 sq ft constructed of platform wood framing with a brick veneer. There was a poured concrete foundation with a finished walkout basement in the rear. The first floor area that sustained the structural failure was comprised of 2 × 10-in. wood joists on 16-in. centers with ¾-in. oriented strand board as the subfloor. The floor was covered with carpet.

The alarm was dispatched to the fire department at 06:11. Engine 102 was the first to arrive on scene at 06:23 hours with a captain, engine operator, and two firefighters. After some confusion and driving past the driveway, Engine 102 laid a 5-in. supply line down the 450-ft driveway to the fire building. Engine 102 was met by the homeowner, stating that the fire was in the basement and everyone was out. Additionally, the excited homeowner directed company members to the front door and indicated where the basement stairway leading to the fire area was located. A 1¾-in. hose line was stretched to the front door in moderate smoke conditions. The report noted that the actions of the homeowner may have distracted the first-arriving crew from doing a proper size-up. After the captain called for water several times, the line was charged and both firefighters took the hose line to the bottom of the stairs but needed additional hose line to advance.

At 06:25, Engine 109 arrived on scene and established Engine 102's water supply at the hydrant on the main road. At 06:26, Ladder 25 with the battalion chief in District Car 25 arrived on scene and assumed incident command. Engine 109's captain informed command that he could not assume accountability because he was busy with establishing the water supply (the hydrant was 500 ft from the entrance drive). Ladder 25 was assigned a search sector and prepared their crew to enter the structure behind Engine 102's interior crew.

At 06:27 the captain radioed, "Engine 102 making entry into the basement, heavy smoke." At 06:30, Engine 109's captain radioed, "Command from Engine 109, contact Engine 102, have them pull out of the first floor, redeploy to the back. It's easy access. Conditions are changing at the front door."

At 06:34 Engine 25 and the designated rapid assistance team had just completed their 360-degree size-up around the building when they encountered one of Engine102's firefighters in front of the building, who reported that he had lost contact with his crew.

The incident commander made several attempts to contact the interior crew (Engine 102) with no response. At 06:37, with the smoke getting black, heavy, and pushing out the front door, the incident commander sent out a "mayday." A rapid intervention team was activated and followed the hoseline through the front door and down to the basement. Returning to the first floor, they noticed a collapsed section of floor and went to investigate the debris in that area of the basement. At 07:08 the captain was found near a corner of the basement. At 07:29, after removing debris from around the captain, the other firefighter was located underneath her and some additional debris. Both victims were pronounced dead at the scene.

The report assumed that the victims exited the basement in heavy smoke and heat by following the hose line and going up the basement stairs to the first floor, where they became disoriented or felt that the quickest escape from the heat was diagonally across the room to an exit door. This path was above the seat of the fire, which had severely damaged the structural integrity of the floor and caused the collapse that pitched them into the basement.

Key contributing factors identified in this investigation included that the initial 360-degree size-up was incomplete, likely disorientation of victims affecting key survival skills, radio communication problems, well-involved basement fire before the department's arrival, and potential fire growth from natural gas utilities.

Risk management analysis. Considering all of these causes, what is the most important factor?

Upon arrival, identifying and prioritizing the on-scene risks must be performed. Life-saving operations is the highest priority and all strategic operations should be directed toward this goal. Priority tactical considerations are the type of structure, the location of the fire, and its severity. This will determine the resources required to confine and control the fire. Immediately following this priority is determining where the fire will be attacked and effective venting.

What were the factors influencing the incomplete 360-degree size-up?

1. The first responding company initially drove past the home, which was up a 450 ft driveway off the main road. I have missed a fire building address while responding and know that there is a tremendous personal pressure on the company officer when any perceived avoidable delay occurs during a response. It creates an atmosphere to hurry and make up the lost time. It is almost certain that the first-due officer in command at this fire was feeling this effect when approaching the fire building.
2. It was not evident from the front that the house was built with a basement on the down slope in the rear, which had a sliding exterior door.
3. A distraught home owner led the company to a stairway near the front door, informing them that this stair went down to the fire location.
4. There was a moderate smoke condition at the time.

It would take a considerable effort by an officer to overcome these circumstances and not be susceptible to the sense of urgency by the owner's insistence to enter the structure immediately. These factors impacted on the officer's ability to perform a proper risk analysis by taking a look in the rear or asking the homeowner if there was another way into the fire area.

Had a size-up included gathering information about another way into the basement, it is likely that the rear door would have been used to attack the fire and a defensive hose line positioned at the front door to prevent fire extension.

To be fair in the judgment of this officer's actions, I admit that there have been times as a company officer when I did not perform an adequate size-up to protect the members under my command or myself when entering a fire building, particularly when there was a known life hazard. Fact. As I gained more experience it became a habit to slow down as I approached a fire building and gather information about the structure. Taking time to perform an adequate size-up should never be ignored by a chief officer. A company officer is engaged in the initial direct actions that will determine if a life is to be saved. A chief officer is one command level away from the actual tactical supervision. Both officers should be performing a risk analysis, but there is a difference.

What other factors affected operations during this fire?

Delayed water supply. The water supply was delayed.

1. The delay in getting water increased the time required for the attack team to begin the interior attack to extinguishing the fire.
2. The second-arriving engine officer was assigned to perform accountability and was diverted from this duty to ensure an adequate water supply as quickly as possible.
3. The limited number of crewmembers on the first-due engine did not allow for personnel to assist in providing an adequate water supply.

If the first-due engine had had an additional firefighter to assist in stretching the hydrant supply line water and pull hose into the fire building, water could have been applied to the seat of the fire in a more rapid manner. Across the country, current fiscal restraints are eliminating this critical additional manpower. Had that initial hose line operated on the fire, the second engine officer would have been able to perform the duties of accounting for who was in the fire building. It might have made a difference at this fire.

Lack of available personnel should be considered a significant risk factor. The resources available on the fireground dictate what tactical procedures can be performed. Resources identified during the initial size-up should be continually evaluated to determine if any changes in strategy need to be made. When water supply does not safely support aggressive interior tactics, this is a priority risk factor and the incident commander should evaluate strategy to control risk.

Radio communications. Radio communications were not effective and hindered a safe operation. In this incident, the engine operator could not initially hear the call for water over the ambient noise of the apparatus. Also, when victim 1 tried to transmit (at least three times) what is believed to have been a Mayday it was not heard or understood.

The report recommendations included having the incident commander assigned an aide to assist in communication capabilities. This is another personnel issue that affects safety.

Survival training. The report stated that basic techniques taught during entry-level firefighting programs describe how to escape a zero-visibility environment using only a hose line and those firefighters may have overlooked this technique. A firefighter operating on a hose line in zero visibility should search along the hose until a coupling is found. Once found, the firefighter can "read" the coupling and determine the male (which will be on the exit side of the connection) and female ends (which will be on the nozzle side of the connection).

I remember being taught this technique as a probie who was glad to be in an engine with a hose line to lead me out and wondered how the ladder company probies would do it. However, when operating at a multiple alarm fire in heavy smoke, anxiously searching for the way out by following the hose line, I was confronted with a knot of twisting, cross looping multiple hose lines that proved extremely difficult to follow in the right direction toward the exit. After that experience, I stopped relying on only one technique and developed skills to follow a wall when possible and note furnishings or structure features in my path. In some cases, following the hose line out is not always as simple as it seems, especially when there are multiple hose lines in heavy smoke and high heat conditions.

Well-involved fire on arrival. There was a well-involved fire on arrival. During the time period between 06:29:24 and 06:34:48, the investigation committee believed that one or more catastrophic events occurred, including a failure of the main-level flooring system. This puts the time from when the fire was reported to the collapse between 18 to 23 min. It was also reported that 23 min from automatic alarm, contact was lost with the victims.[66] The area of fire origin had no finished ceiling other than the suspended acoustic tiles. The undersides of the first floor joists were quickly exposed to direct fire impingement, causing rapid deterioration and failure of the flooring system directly underneath the first floor level.

The American Society for Testing and Materials Assembly Test indicated that a traditional 2 × 10-in. structural member failed in 12 min, 6 sec.[67]

It was suspected that this fire was gas-fed, which contributed to the intensity of the fire and resulted in weakening the load-bearing members. Shutting down the utilities almost always improves conditions inside the fire building and should be given a priority consideration, especially during interior operations.

Conclusions. Every routine residential fire has the potential for several things to go wrong at once. A perfect storm of negative occurrences will happen when there is an out-of-control fire inside a building. When things are not going right from the beginning and essential tasks are being hampered or delayed, the incident commander should recognize these signs and take actions to control risk by suspending, altering, or terminating operations where necessary.

Firefighter fatality investigation 6

A fire occurred at 6:56 p.m. on June 18, 2007, in the Sofa Super Store in Charleston, South Carolina. The store manager located the fire inside the loading dock and upon returning to the showroom area, asked other employees to call 911 and to exit the store through the front doors of the showroom. At 7:08 p.m., the report of the fire was received by the Charleston County 911 Emergency Center and the Charleston Fire Department was dispatched.[68]

Response. The fire department arrived on scene in fewer than 4 minutes after the 911 dispatch received the report of an exterior trash fire behind the Sofa Super Store. Upon initial survey of the exterior of the structure, the fire department located a fire inside the enclosed loading dock. A recon of the interior did not find any smoke or fire in the showrooms. Three minutes after arrival, a recon of the rear of the west showroom confirmed the loading area was fully involved in fire.

By 7:12 p.m., additional units, including an assistant chief and Engine 11, had arrived on scene. When the assistant chief asked employees if there was anyone left in the store, the response was that everybody was out.

By 7:16 p.m., fire crews were applying water using a 1½-in. diameter hose line on the fire in the loading dock. At the same time, at the front of the store, crews were pulling a 1½-in. diameter hose line in the front door, through the main showroom, and into the west showroom.

The fire chief had arrived on scene and became the incident commander. The fire chief met with the assistant chief and battalion chief 4 and instructed the assistant chief to direct operations at the front of the store while the fire chief would direct loading dock operations. The fire chief subsequently directed operations at the loading dock, but also functioned as the incident commander by requesting additional resources, coordinating coverage of the city when units responded to the fire, and requesting the water company to increase water pressure.

Upon arrival in the front of the store, the assistant chief directed Engine 16 to stretch a 2½-in. hose in the front door. Two hose lines were now stretched through the front doors and pulled to the rear of the west showroom. While fire crews were pulling the hoses to the rear of the west showroom, other companies worked on suppressing the fire on the loading dock. They did not know that the fire had spread through the open roll-up door into the holding area, then into the void space above the drop ceiling, and eventually into the rear enclosed ceiling space of the main showroom.

At approximately 7:27 p.m., dispatch notified the fire chief of a cell phone call from a man claiming to be trapped inside the store. The caller indicated to the fire department that he was an employee who worked in the repair shop. Another roll-up fire door had closed and prevented the employee from escaping the warehouse/repair shop area. The employee was banging on the exterior metal wall with a hammer. The fire chief radioed the assistant chief and directed him to respond to the trapped employee. At this stage in the fire, all firefighters and chiefs were using the same radio channel, so all firefighters with a radio heard the report of a trapped employee in the rear of the store. The assistant chief asked the store manager if the trapped person was an employee of the store, and the store staff verified that the man was someone who repaired furniture for the store. The assistant chief took a team of firefighters and went around the east end of the store, chopped through a locked wooden gate, and located the employee banging on the metal wall. Using pry bars, the firefighters were able to create an opening in the metal wall and extract the trapped employee. At approximately 7:31 p.m., the assistant chief, rescue team, and rescued employee returned to the front of the store.

During the rescue effort at about 7:27 p.m., several inaudible radio communications suggested that someone was trapped inside. It was not clear whether the calls reported that firefighters were lost or trapped or whether the calls were related to the trapped employee.

Several minutes later, between 7:29 p.m. and 7:30 p.m., there were additional radio communications that were still difficult to understand, but it seemed that one or more firefighters were asking for directions to exit or requesting assistance to escape. Other radio calls were interspersed with calls for help related to getting the trapped employee out.

Beginning around 7:31 p.m., additional broken radio traffic clearly indicated that several firefighters were in distress. An unknown firefighter called "Mayday" and dispatch advised the fire chief that the L-5 engineer had activated the emergency button on his radio.

The fire chief radioed, ". . . we need to vacate the building," but it is not clear what was done at the front of the store to implement this order. Apparatus air horns were sounded in a rapid manner to alert members to vacate the building. Conditions within the store deteriorated rapidly. At the loading dock, the fire chief held back fire teams from reentering.

By the time the assistant chief returned to the front of the store from rescuing the trapped employee, fire conditions had changed dramatically. Brown smoke was rolling out of the front doors. Radio communications were indicating that firefighters had become disoriented or lost within the store.

Inside the showroom were crewmembers of Engines 6, 11, 15, 16, and 19 and L-5. At this stage in operations, the crews had already stretched a 1½-in. hose and a 2½-in. hose through the front doors, through the main showroom, and to the rear of the west showroom. Also, another 2½-in. hose had been stretched about 75 ft into the main showroom. All three lines were operating, but the amount of water being supplied was not sufficient to control the fire at the rear of the main showroom. It appeared that Engine 11 was not able to provide sufficient water to supply all three hose lines simultaneously.

As the fire continued to grow in the rear of the main showroom, the volume of smoke thickened around the firefighters and eventually dropped closer to the floor. As the visibility decreased, the firefighters within the smoke-filled main showroom became disoriented as evidenced by firefighters requesting help by radio transmissions.

At about 7:35 p.m., the front windows of the main showroom were broken out; immediately heavy brown smoke poured from the broken windows. Less than 1 min later, the smoke changed to thick black smoke. Then flames emerged from the front windows within 3 min of the windows being vented. The fire spread rapidly from the rear of the main showroom to the front and out the front windows of the main showroom. The fire also spread rapidly into the west showroom through three roll-up fire doors that did not close.[69]

Nine firefighters were trapped in the fire, resulting in their deaths.

Intense heat from the burning furniture in the main showroom weakened the roof joists and supports and resulted in the collapse of a portion of the roof over the main showroom at approximately 7:51 p.m., 13 min after flames emerged from the front windows (40 min after the fire department arrived on scene). Portions of the parapet wall and front façade also collapsed onto the front parking lot.

The fire continued to burn vigorously in the main showroom and eventually almost the entire roof over the main showroom collapsed onto the showroom floor. Merchandise in the showrooms and warehouse continued to burn for an additional 2 hours and 20 minutes before the fire was extinguished.[70]

Risk management analysis.

- Risk identification: Risk identification is the first step in evaluating the incident and begins with the initial response plan.
- Failure of CFD inspections to identify hazards: CFD inspections did not identify the large fuel load, the non-code compliant wood construction, the solvent storage on the loading dock, or the lack of a fire door between loading dock and holding area as significant fire hazards. The inspections of the structure did record a lack of fire alarms and sprinklers.
- Staffing: The initial response to the Sofa Super Store was, as per CFD procedures for fires involving structures less than five stories in height, a first alarm assignment of two engine companies, a ladder truck company, and a battalion chief. With an engineer, a firefighter, and an officer on each apparatus, the fire department's initial response included 10 people.

According to *NFPA 1500*, the Sofa Super Store building would likely have been identified as high hazard occupancy due to the lack of sprinklers and the presence of large open areas and a large fuel load. The response for a high hazard occupancy, based on *NFPA 1710*, advocates a minimum crew size of 5 to 6 members for each apparatus, which for this incident would amount to 16 to 19 people for the initial response.

At the time of the fire, the CFD's unit staffing was less than the minimum complement of engine and truck company personnel recommended by the National Fire Protection Association. *NFPA 1500: Standard on Fire Department Occupational Safety and Health Program* requires that a fire department develop and adopt a comprehensive risk management plan to identify and evaluate potential building/occupancy hazards. Based on the building's hazard designation, recommendations are made regarding staffing of responding.[71]

Staffing levels directly affect the ability of firefighters to perform basic tactical operations. Staffing levels determine how effectively a department can control a fire or mitigate an emergency based on how long it takes to establish water supply, advance hose lines, effect rescues, and perform several tactical procedures which are mandated for various situations.

At this fire, staffing issues impacted on these operations:

1. Adequate water supply required to supply several hose lines simultaneously is critical to provide safety to operating forces and control a large area fire. Had there been available staffing to quickly stretch hose lines and position apparatus, an effective water supply could have been established, which might have made a significant difference at this fire.
2. The CFD did not ventilate the roof in the rear of the main showroom. There were no doors or windows on the rear (south) side of the main showroom. Roof ventilation could have allowed the heated smoke gases to lift out of the building, creating a tenable condition for the firefighters inside and allowing them to exit the structure quickly or possibly contain and/or extinguish the fire. Ventilation timing is critical. Actions by the personnel venting and those attacking the fire must be tightly coordinated by the incident commander. Proper roof ventilation of fires inside buildings is a proven method of increasing the potential to save lives and improving firefighting tactical operations.
3. The fire area involved several interconnected large buildings, which required areas to be broken down into sectors with chief officers supervising personnel in each sector. That staffing was not available to the incident commander. At this fire the fire chief was supervising the initial fire operations in the loading dock area while directing incoming units and various other incident command responsibilities. The assistant

chief had to be in two places at once; directing a confirmed life-saving operation in one area and responsible for supervising the front of the building where units were engaged in operations at the early stages of a deadly fire.

4. At large-scale operations such as this, one command post needs to be established and staffed with support personnel to help provide firefighter accountability, monitor communications, and assist the incident commander with overall command and control of the incident.
5. CFD procedures allow off-duty firefighters to respond to fires and emergencies. Upon arrival at the incident, department procedures require each firefighter to provide a chief officer with an identification card before operating at the scene. They do not require the firefighter to report into the incident commander specifically. Off-duty firefighters checking in to different chief officers hinder accountability. At this fire it was not clear who was directing operations at the front of the store or inside the showrooms while the assistant chief was leading the life-saving rescue efforts.

Mutual aid provides most of the staffing for fire department operations at large-scale incidents. Mutual aid agreements vary; some require the incident commander to specifically request particular units, while others require routine deployment of another department's units to the scene. These agreements benefit all participants by providing staffing and equipment that would be too expensive to maintain in each separate jurisdiction.

Generally, mutual aid effectiveness is hampered by differences in tactics, communications, and equipment. It is the mark of a professional, effective organization when resources can be put in place on demand in various conditions and locations and still function at a high level. The U.S. National Wildfire Coordinating Group has this capability and is a good example of how an effective incident command system works. The challenges to create effective mutual aid arrangements that cover a multicounty or a multistate region are not to be underestimated. A commitment by all those involved—state and local authorities and fire department leaders and members—is necessary to create a response plan that is the best possible solution to mutual aid problems. In many ways this is one of the most challenging issues facing fire departments across the country.

Risk evaluation. The responders did not know when fire and smoke entered into the space between the dropped ceilings and the roof above the showrooms. The NIST study was not able to document if any firefighter removed a ceiling tile to check for fire spread above the drop ceiling. There were no heat or smoke alarms installed in the building and there was no central fire alarm panel to provide information to the firefighters.[72]

Brown smoke color is consistent with large amounts of partially burned combustion products from a fire that does not have sufficient air for complete combustion. As this fire continued to grow and spread in the ceiling space of the main showroom, brown smoke began to emerge from the front doors in increasing quantities.

After the windows were vented, the smoke changed color and became much blacker. The air provided additional oxygen to the fire, which increased and released more energy. As the fire at the rear of the main showroom ignited the smoke layer of unburned combustion products, the fire rapidly moved from the rear of the showroom to the front of the store.

Fire Chronicle

One of the early fires in my career involved a five-story, brick walls with wooden floor joists, commercial building. As we approached the building I could see heavy brown smoke whirling around behind three windows on the second floor with the windows actually moving in and out as if the building was breathing. Moments later these windows blew out and a ball of fire rolled into the night sky. After that, anytime there was heavy, thick, rapid smoke movement confined inside a building I knew that a hazardous backdraft/smoke explosion condition existed.

* * * *

The combination of the color of the smoke (which can be a mixture of black, brown, and greenish yellow) and movement indicates heated gases under pressure, which can violently ignite if an air source is provided. These are conditions that should be evaluated to determine a risk/benefit analysis of current fireground operations.

Compartmentation. The roll-up fire door in the breezeway between the holding area and warehouse had activated and closed direct access to the warehouse. The fire inside the enclosed loading dock spread to the front of the warehouse through a shared corrugated metal wall. The fire heated the metal wall enough to cause items inside the warehouse to ignite.

Postfire inspection of the holding room documented that the roll-up door from the loading dock to the holding area did not close during the fire. Examination of the operating mechanism after the fire demonstrated that this roll-up door did not have fusible links or other components that would have automatically closed the door in case of a fire. The smoke and flames from the loading dock flowed through this open door and into the holding area where the thermal radiation from smoke and flames ignited furniture.[73] A *risk analysis* of the large open area with a lack of compartmentation, no sprinklers, a heavy fire load, combined with the type and color of smoke conditions, would be classified as a "low frequency/high severity" condition requiring a complete reevaluation of the ongoing strategies and tactics being deployed by the incident commander. Unfortunately at this fire there was a lack of information regarding the complex interconnected buildings, a lack of effective building codes providing compartmentation and sprinkler protection, and there were insufficient chief officers available to cover multiple incidents occurring simultaneously.

Some of the NIST recommendations that were generated by this investigation are:

- Recommendation 1: adopt building and fire codes based on the model codes. Current model codes applied retroactively to high fuel-load mercantile occupancies would have required a sprinkler system to be installed.
- Recommendation 2: implement aggressive and effective fire inspections/enforcement procedures.
- Recommendation 5: develop comprehensive risk management plans that identify hazards and allocate resources according to identified risk.[74]

Once again, fire prevention is noted as playing a major role in protecting occupants and first responders.

Prefire Planning

Prefire plans will assist in providing risk identification, which will alert the first responders of the dangers involved and how to operate effectively in a particular environment.

Fire departments must allocate time and resources to prefire plan structures in their response area. Using common sense can put this seemingly huge task into proportion. Start with large buildings or any structure that requires a significant response to a fire or emergency. Once data has been completed for these structures, smaller buildings and one- or two-family homes can be addressed.

NFPA 1500 is one of many of the many publications related to creating prefire plans. Use publications like this as a guide to make the prefire plans suit your department needs. The main thing is not to become overwhelmed by complex building features. Keep it simple. Here is a brief layout regarding what general information should be collected for a prefire/emergency plan.

1. What type of occupancy exists? Is it a high, medium, or low hazard?
2. What type of fire protection exists? Are there sprinklers or standpipes, does the building have smoke detectors with an alarm panel, and are major load-bearing structural members protected or exposed?
3. Define any special hazards at the location; that is, dangerous chemicals, radioactive materials, and other types of toxic substances; electrical hazardous; and any locations where a significant threat to the occupants and first responders is a consideration.

Once this information is collected, begin working on operating procedures that address the apparent risks. Providing appropriately trained firefighting staffing will be a factor in controlling fires and emergencies in many hazardous occupancies. This reality should be brought to the attention of local officials and the safety issues to occupants and firefighters clearly explained. The facts can sometimes prevail toward gathering resources needed for required staffing to mitigate specific threats to life and property in an appropriate manner.

Some features of a prefire plan are unavoidable; it takes time and money to get it done. One of the main obstructions to creating effective prefire plans is that there are often structural alterations to buildings that are difficult to track.

Another major difficulty is to record information and update existing building data so that it is accurate for use by emergency responders. As they say, the devil is in the details.

Which type alteration should be considered for inclusion into the prefire plan? Plumbing a new bathroom? Electrical wiring a new floor for heavy technology use? Building a partition wall in an office space? Raising a ceiling?

In order to determine the answer to what type of building alterations should be included in drawings used at emergency operations, it is required to have a group of personnel dedicated to this task working with the building department and fire department. It requires significant effort to establish interagency communications needed to capture and exchange that information and update existing prefire plans to reflect these types of changes.

In addition to these actions, laws must be passed requiring building owners to submit their existing and/or updated build plans. Without stringent legal requirements, the information will not be supplied.

Once all of these tasks have been completed, another limitation must be recognized and that is the ability to capture accurate, complete structural information on *existing structures*. Research of building records is required to find out what work was done and ascertain if the work performed was approved. In older buildings this information is often not available.

Existing altered structural conditions are often difficult to find due to inaccessible locations, and/or having major structural elements covered over with sheetrock or other material.

Even if conditions are visible in existing buildings, alterations involving structural connection points frequently require qualified structural engineers to recognize the load forces being applied as a result of the alteration and the stresses being taken on by other structural elements caused by the new loads. There are significant challenges in how all this data is stored and then presented in a manner that can be effectively used by the incident commander. Technology must be able to support the information retrieval and present it in a comprehensive manner on the fireground. All of this is completely worthwhile to do for the citizens and firefighters of this country, but hard to accomplish without local government resources.

Overview of firefighter fatalities

A study performed on trends in firefighter fatalities due to structural collapse from 1979 to 2002 reported that there were 63 deaths caused by structural collapse in a total of 47 fires.[75] The majority of collapse fatalities occurred during fire attack. The cause of these deaths was attributed as follows:

- In 60% of deaths, the cause was a firefighter being caught or trapped inside a structure and running out of air supply or succumbing to injuries.
- In 40% of deaths, the cause was a firefighter being struck and severely injured by a part of collapsing structural material.

While the report noted that on an annual basis, collapse fatalities have declined since 1979, it did mention that the percentage of collapse fatalities that occurred in residential properties has increased.[76]

There was no information given for the reasons of the increase in fatalities in residential structures, but it is likely that lightweight construction is a major reason for these structural failures. The fire departments have a history and a culture that is common throughout the world. We are proud of that history and know that it will continue, some of it at a terrible price. A memorial on the wall of the quarters of FDNY's Engine 10 and Ladder 10 recognizes this spirit with these words: "Dedicated to those who fell and to those who carry on."

Building Codes and the Fire Service

It wasn't until I had over 30 years of active field firefighting duties before I was asked by the Chief of Department Pete Hayden to run the Bureau of Fire Prevention for the FDNY. It is not considered a glorious job. As a matter of fact it is one of the least desirable positions for most career firefighters. There is no battle glory in writing building and fire codes. What is recognized by the leaders of fire departments is that fire prevention is the MOST effective way to protect life and property. As I began meeting colleagues in this discipline I found that building and fire codes are, for the most part, decided on by people who have never been inside a building on fire. In most states, the fire code official does not come from the ranks of active firefighters. The NYC Bureau of Fire Prevention is run by a small group of uniformed members with civilian personnel supervising most of the activities and performing the day-to-day duties and inspections of the bureau. Uniformed firefighters perform scheduled fire prevention limited inspections in their respective local administrative areas.

During my tenure as chief of fire prevention, I discovered the importance of being involved in fire and building code issues. Any code written to protect the lives of civilians will also protect firefighters.

Requiring sprinklers in one- and two-family homes will decrease the number of lives lost and reduce the incidents of fire-induced building collapse in these types of structures. The issue preventing this code requirement is, of course, cost. Fire service personnel need to work closely with building professionals to find ways to keep sprinkler installation costs to a minimum in one- and two-family homes in order to realize this goal.

Sprinkler activation is an efficient fire protection system, but what can occur when a sprinkler system fails must also be considered. When the sprinkler systems fail to limit fire spread and it grows beyond the capabilities of extinguishment by firefighters, the only system that keeps a structure from collapsing is the fire-resistant design of the building. Buildings should not be left to burn to the ground because the sprinklers fail to operate and the fire grows to the point where firefighting efforts cannot extinguish it. Passive, fire-resistant building design features such as fire separations, enclosing structural members with noncombustible materials, and compartmentation will limit fire growth and determine the loss of life and property when active firefighting measures are not effective.

Architects and engineers do not design buildings to withstand the nominal load forces of fire damage. It simply is not a consideration. An accurate understanding of how structural elements react to fire damage and ways to counter these forces will give these building professionals the knowledge they need to design structures resistant to fire-induced collapse. Designing buildings to address this probability is something that should be done. Fires can start small and develop into major conflagrations with a speed that often astonishes inexperienced observers. These types of events will continue to happen due to the "unforeseen consequences" of daily life.

How well a building resisted fire damage, when comparing relative similar building fires, can be determined by a simple criteria: How soon after the fire were the occupants able to use the building as it was intended to be used? Was it hours, days, months, or never? Structural weakness is the prime concern that must be addressed before allowing occupants to re-enter a fire building. The longer building occupants are kept out of the building, the higher the loss of revenue.

Building professionals and insurance companies need to become aware of the injuries and/or fatalities incurred and the dollar loss from structural fire damage if there is going to be support for spending more money on a structure's fire-resistant capabilities.

The number of partial and/or complete fire-induced structural collapses is grossly underestimated. The amount of money lost due to structural collapse is not defined separately from the total lost dollar figure for each individual building. The bottom line figures often do not calculate the amount of time it took for the building to return to its original, revenue-generating intended use. Also it is very difficult to capture the value of lives and money saved when active and passive fire protection systems function properly. If these calculations were defined, the dollar numbers would be staggering. Fire departments have a responsibility to capture information when a fire causes partial or complete collapse of a structure. This should be a matter of routine. As stated earlier, the NIFRS system of recording fire events needs to be modified so that it requires recording collapse information, even when it involves partial or local collapse such as a ceiling failing in a 10 × 10 ft room. The collapse information captured should include type of construction, phase of construction (initial construction, alteration, or demolition), size and location of the collapse area, the type of structural materials involved, and the effect the collapse had on other structural elements. Routinely documenting the collapse event and including the pictures in the report would greatly assist in using the information for research purposes. There are computer systems available to collect this data, making it a reasonable request to download pictures during the report writing process.

Making this type of information available to architects, building owners, and insurance companies would help them recognize the high cost of structural failure. Firefighters' firsthand knowledge of these events must be communicated to the code officials and the construction industry.

Prescriptive versus performance code

A recent change in the way building laws are being interpreted is challenging the ability of fire inspectors to determine the reliability of fire-resistant construction.

Originally, building codes were based on a *prescriptive code* (also known as a *specification code*), which is a building code that specifies construction requirements according to particular materials and construction methods.

Now it is becoming common to use a *performance* type of building code that specifies construction requirements according to performance criteria rather than to specific building materials, products, and/or methods of construction. The difference between these two codes can be explained as follows: A code would be considered as *performance* code if it requires the completed work to satisfy specified standards such as a wall must be able to withstand 3 hours of fire impingement and remain stable, without describing in detail how to satisfy those standards. A *prescriptive* code would require certain materials to be used such as the wall must be constructed of masonry material and, to some degree, describe how it must be built.

Performance codes allow the builder to use any combination of materials and methods that will provide the fire resistance necessary to satisfy the code. This type of code makes available new improved materials and techniques that satisfy building code standards. However, this use of various methods and materials makes performance codes more difficult to enforce. Code officials must constantly evaluate new techniques and fire-engineered products that often use computer models and limited fire-resistance test results as proof of performance. On one hand it is appropriate for architects, engineers, and the construction industry to use available new products to keep costs down and improve building design versatility. What goes hand and hand with this is that code officials are now, for the most part, taken out of the decision-making process regarding what products will provide appropriate fire protection properties other than establishing hourly ratings. With performance codes, officials must rely almost completely on industry standards to determine what construction products are appropriate to be used. The concern is that this system relies totally on testing standards that often do not provide information that accurately defines a material's or building system's reaction to fire damage. With *prescription codes* there are more areas of direct input from code officials that ensure an element of safety that does not occur with *performance codes*. Why? Because with *performance codes* it takes a qualified staff dedicated to reviewing and analyzing products and systems to determine their respective fire protection characteristics. The time required and the cost to hire this type of staff is simply not feasible for most fire department fire prevention bureau budgets.

Finding ways to share information that protects the public and our firefighting personnel is a struggle. However it has been my experience that most fire departments welcome communications from other professions. These communications must include fire prevention solutions that save lives and protect property.

Who can the fire service rely on?

I had a friend whose brother worked in NYC's Mayor Koch's administration when they were discussing emergency plans in the event city workers went on strike. He told me the details. They had a plan for the sanitation department; contracts were drawn up to provide garbage pickups by private contractors. The National Guard call-up procedures were reviewed for temporary replacement of the police department. Next was the fire department. They looked around the table at each other and no one had a suggestion as to who would replace firefighters.

There is no one to come and rescue firefighters other than themselves. We rely on each other to survive the dangers we face. The same is true when it comes to creating building and fire codes that can help prevent the collapse of a burning building and save firefighters lives. We must do it ourselves.

I have met with many fire protection engineers and they have been wholeheartedly supportive of firefighters. They joined me in my efforts when I proposed codes that were being fought by the real-estate interest and building owners due to increased costs. I welcomed their involvement and benefitted from their support. The fire protection engineers that I have met impressed me with their sincerity to provide the best possible safe building codes. However, fire engineers are presumed to have a thorough understanding of the behavior of an out-of-control fire inside a structure without the benefit of actually witnessing a real fire. They are not required to have any firefighting experience, nor are they required to be a member of the fire service.

I believe their profession would welcome an opportunity to include more interactions with firefighters. A solution to their lack of experience with actual fire conditions might be to require a certain amount of time spent witnessing fireground operations. This could be accomplished by creating an agreement with local fire departments to spend time riding with first responders or to respond to the scene of a fire and be included in the critiques of fire operations. Another alternative could be to require service in either a paid or volunteer fire department as part of their formal education requirements.

Fire engineers need to recognize the limitations of computer modeling due to a lack of accurate data on fire-induced building collapse and place more reliance on proven fire-resistant materials and codes. I know of many fire engineers who support this balanced approach to fire safety and want codes that reflect these concerns.

Firefighters must be present during code conferences such as those held by the International Code Council and NFPA code meetings. There are many construction industry lobbyists who take a very active role in affecting the way a fire protection code will be written. New products are constantly being introduced that have a significant impact on a building's fire protection. Fire department personnel must recognize this reality and become involved in lobbying to have safe codes enacted into their local legislation. This means becoming involved in the political process. There is no reason to be totally adverse to the legal process. For the most part I found local politicians willing to support issues that promote the safety of their constituents. If accurate information regarding the dollar loss from fire-induced damage and/or collapse becomes available, it can be compared to the additional cost of adequate fire protection. I believe that this type of analysis would show that when a fire occurs within a structure, the financial bottom line could support spending the money on construction for adequate passive and active fire protection to facilitate a more rapid return to generating the revenues intended from the original investment in the building.

Agencies that provide standards for building and fire codes supply a valuable service and their standards are high in most of their testing methods. However they generally answer only to the client who establishes the parameters on what a product will be tested for. Often, there is no obligation to find data on the final failure temperature of a particular structural product. Any additional information beyond the standard minimum test is a cost factor that needs to be justified. No villains here, just the system.

The standards define minimum fire-resistant characteristics and the products are designed to meet related temperatures as per code requirements. However, this is a problem when trying to find information regarding fire-induced building collapse. If the failure temperatures of the involved building materials are not known, the search for information hits a dead end.

The testing procedures for fire-resistant materials must to be changed in order to obtain accurate information that can be used during the building design process. NIST, a FEMA agency under the Department of Homeland Security, sounded the alarm of concerned fire chiefs regarding the inadequate testing methods. NIST deserves credit for making the effort and accurately determining what needs to be changed in fire-resistant testing methods and offering suggestions on how a building can be built to resist fire-induced collapse.

It is the responsibility of local fire and building code personnel to take the information provided by NIST to create substantial changes in fire protection standards and codes. The building code writers are responsible for mandating safety regulations for their jurisdictions. All firefighters must support them in accomplishing this goal.

Each mayor wants his or her citizens to have construction projects that provide jobs and increase revenues. If a builder complains that the state or local construction codes are costing too much and that this additional cost will force the builder to seek another location for the project, the elected officials might try to accommodate the enterprise by putting pressure on the code officials to fall in line with what is accepted elsewhere if that cost is less.

Fire and building code officials have to educate the public and politicians on the reason for the added cost that always comes with an increase in fire protection. The truth of the matter is that America has some of the strictest building codes found anywhere in the world. We know how to build it and we build it right. No one can build better buildings than America's construction industry. It is truly something to be proud of and we need to continue to support actions that will ensure our status as the leaders in building safe buildings.

Battles to build safer buildings are going on daily, but rarely by active firefighters. Firefighters need to be a part of this process because they are the ones who have firsthand knowledge of the consequences to life and property when a building collapses as a result of an out-of-control fire.

This can only help in creating the best structures for the citizens we serve and decrease the number of fire-induced building collapses.

References

1. Andersen, Kurt. "From Mao to Wow!" *Vanity Fair*, August 2008, 576.
2. Zoninsein, Manuela."New Ideas Surface About TVCC Fire in Beijing," *Architectural Record*, March 13, 2009.
3. OKC Jeff. "Enormous fire kills one and burns brand-new CCTV building in Beijing," Hubpages. http://okcjeff.hubpages.com/hub/Beijing-CCTV-Tower-Fire.
4. Blanchard, Ben. "Fire Claims Building at CCTV Beijing Headquarters," FirefighterNation, Feb 9, 2009. http://my.firefighternation.com/forum/topics/fire-claims-building-at-cctv.
5. "Funding Comes Alight for Emergency Service," China Daily Information Company, Jan 26, 2010, http://www.chinadaily.com.cn/metro/2010-01/26/content_9376904.htm.
6. Andersen.
7. The Concrete Centre. "Madrid Windsor Tower Building Fire, 14–15 February 2005," case study. Retrieved from www.concretecentre.com, November 2011.
8. Ibid.
9. Beitel, Jesse and Nestor Iwankiw. "Analysis of Needs and Existing Capabilities for Full-Scale Fire Resistance Testing," National Institute of Standards and Technology (NIST), NIST GCR 02-843-1, Section 2.3.1, October 2008.
10. Ibid.
11. Ibid.
12. Ibid.
13. Gann, R.G. "Final Report on the Collapse of World Trade Center Building 7, Federal Building and Fire Safety Investigation of the World Trade Center Disaster (NIST NCSTAR 1A)." Retrieved from www.nist.gov November 2011.
14. Ibid.
15. Ibid.
16. Ibid.

17. American Society of Civil Engineers. "ASCE Report on World Trade Center Collapse Released; House Committee Holds Hearing on Findings," for the week ending May 3, 2002. Retrieved from www.asce.org, November 2011.
18. Beitel.
19. Ibid.
20. Hashagen, Paul. *FDNY: The Bravest, An Illustrated History 1865–2002*. Hudson, MA: FSP Books & Videos, 2002
21. *FDNY Fire Prevention Manual*, Chapter 12, "Fire Prevention Information Bulletin 30."
22. Chasteauneuf, Paul. "A Guide to Testing Stairwell and Passage Pressurization Systems to AS/NZS 1668 pt-1 1998." *The Official Journal of AIRAH*, September 2002.
23. Jennings, Charles. "An Effectiveness Comparison of Sprinklers and Compartmentation for High Rise Office Building Fire Protection as Defined by Local Law 5 (1973) for the Years 1981–1985," 1990 Masters thesis, John Jay College of Criminal Justice of the City University of New York.
24. "Fire Prevention Information Bulletin 30." *FDNY Fire Prevention Manual*, chapter 12.
25. Robinson, Henry Morton. "Killing Fires High in the Air," *Popular Science Monthly*, December 1928.
26. Ibid.
27. Ibid.
28. The American Society of Civil Engineers. "1.4, General Structural Integrity," *Minimum Design Loads for Buildings and Other Structures*, 2005.
29. The American Concrete Institute. "Reinforcing Steel Details," *ACI 318-02: Building Code Requirements for Structural Concrete*.
30. Nair, Shankar. "Preventing Disproportionate Collapse," *Journal of Performance of Constructed Facilities*, Volume 20, Issue 4, 2006.
31. Public Buildings Service (PBS) of the U.S. General Services Administration (GSA). *Facilities Standards for the Public Buildings Service*, 2000.
32. Public Buildings Service (PBS) of the U.S. General Services Administration (GSA). *Facilities Standards for the Public Buildings Service*, 2003.
33. The American Concrete Institute. "Structural Integrity: Reinforcing Steel Details," *ACI 318-02: Building Code Requirements for Structural Concrete*.
34. FDNY Firefighting procedures, highrise buildings.
35. National Fire Protection Association. *NFPA 14: Standard for the Installation of Standpipe and Hose Systems*, 2010.

36. Hill, Howard J. "High Pressure Trilogy—The Drill at 7 World Trade Center," *WNYF* 2007, Number 3.
37. Dunn, Vincent. *Safety and Survival on the Fireground*. Saddle Brook, NJ: Fire Engineering Books & Videos, 1992.
38. United States Fire Administration National Fire Data Center. "New York City Bank Building Fire: Compartmentation vs. Sprinklers," January 31, 1993.
39. Fire Engineering (August 1959) and Iowa State University. Bulletin 18, "Water for Fire Fighting: Rate-of-Flow Formula," 1959.
40. National Institute for Occupational Safety and Health (NIOSH). "Restaurant Fire Claims the Life of Two Career Fire Fighters—Texas," *F 2000-13 Fire Fighter Fatality Investigation and Prevention Program Summary*. February 7, 2001.
41. Ibid.
42. Ibid.
43. ASTM International (formerly American Society for Testing and Materials). *ASTM E119*.
44. Naum, Christopher J., SFPE. "National Fire Fighter Near-Miss Reporting System May-Structural Collapse." Report sponsored by the National Fallen Firefighters Foundation and the U.S. Fire Administration, May 2009.
45. Murphy, Bill. "'Fast attack Questioned in Houston Fire Death," *Houston Chronicle*, February 19, 2006.
46. "Battalion Chief and Career Master Fire Fighter Die and Twenty-Nine Career Fire Fighters are Injured during a Five Alarm Church Fire—Pennsylvania," National Institute for Occupational Safety and Health (NIOSH), *F2004-17 Fire Fighter Fatality Investigation and Prevention Program Summary*, January 27, 2006.
47. Ibid.
48. Ibid.
49. Ibid.
50. Ibid.
51. Roosevelt, Theodore, "Citizenship in a Republic," speech at the Sorbonne, Paris, April 23, 1910.
52. "Four Firefighters Die in Seattle Warehouse Fire, Seattle, Washington." United States Fire Administration National Fire Data Center, USFA-TR-077/January 1995.
53. Ibid.
54. Ibid.
55. Ibid.

56. Ibid.
57. Ibid.
58. Ibid.
59. Lowery, Robert O., Fire Commissioner. "Board of Inquiry Report of 23rd Street Fire and Building Collapse," October 17, 1966.
60. Hill, Howard J. "Revisiting the 23rd Street Fire," *WNYF*, 2006, Number 3.
61. Lowery. "Board of Inquiry Report of 23rd Street Fire and Building Collapse."
62. Ibid.
63. Ibid.
64. Hill. "Revisiting the 23rd Street Fire."
65. Colerain Township Department of Fire and Emergency Medical Services, Ohio. "Investigation Analysis of the Squirrels Nest Lane Firefighter Line of Duty Deaths," April 4, 2008.
66. Ibid.
67. *ASTM E119*.
68. Bryner, N.P., S.P. Fuss, B.W. Klein, A.D. Putorti Jr. "Technical Study of the Sofa Super Store Fire—South Carolina, June 18, 2007," *NIST–SP 1119 Volume II*.
69. Ibid.
70. Ibid.
71. National Fire Protection Association. *NFPA 1500: Standard on Fire Department Occupational Safety and Health Program*, 2007.
72. Bryner, Fuss, Klein, and Putorti.
73. Ibid.
74. Ibid.
75. Brassell, Lori D. and David D. Evans. "Trends in Firefighter Fatalities Due to Structural Collapse 1979–2002," *NISTIR 7069*, November 2003.
76. Ibid.

Index

C

D

E

F

G

H

I

J

K

L

O

P

R

S

T

U

V

W